Breath Taken

THE LANDSCAPE AND BIOGRAPHY OF ASBESTOS

an exhibition by

BILL RAVANESI

Center for Visual Arts in the Public Interest

This catalogue was made possible by
a generous grant from the

Asbestos Victims Special Fund Trust

Mary Vogel, Executive Director

Contents

Catalogue edited by Bill Ravanesi

Center for Visual Arts in the Public Interest, Inc.
348 Congress Street
Boston, MA 02210
(617) 482-8899

The Center for Visual Arts in the Public Interest, Inc. is a 501(c)(3) non-profit organization, whose primary purpose is to promote public awareness of health issues through the visual arts.

Library of Congress Catalogue Card Number 91-071047
ISBN 1-879842-91-2

First Edition

Vintage photographs on inside front cover, inside back cover, pages 3, 8, and 14 from Johns-Manville Archive. Photograph on page 8 by Neal Ranauro.

Designer Margaret Sartor
Printer Franklin Book Press, Providence, Rhode Island

The 36-minute video, "Breath Taken," can be obtained by contacting Fanlight Productions, 47 Halifax Street, Boston, MA 02130 (617) 542-0980.

The exhibition, *Breath Taken: The Landscape and Biography of Asbestos,* is currently travelling nationally under the auspices of the Center for Visual Arts in the Public Interest, Inc. For further information on the national tour schedule, please contact the Center.

Exhibition Itinerary includes:

Boston University Art Gallery
Breath Taken: The Landscape and Biography of Asbestos
March 1-April 8, 1990

Houston Center for Photography
May 18-June 24, 1990

Museo Italo-Americano
Fort Mason Art Center
San Francisco, CA
February-March 1992

Rutgers University
Labor Center
Walters Gallery
New Brunswick, NJ
Date Pending

Marian Locks Gallery
Philadelphia, PA
Date pending

Acknowledgements

I would like to express my deep gratitude for the understanding and friendship extended to me for so many years by the victims, and their families. Their shared experiences have been central to this project.

During the years I have received help and encouragement from so many people that it is not possible to name them all here. I would like to express my deepest appreciation to my mother, Lucy, and thanks to the following people for their time, thoughts, and support so generously given: Anton and Francoise Nelessen, Mary Vogel, Natatcha Estebanez, Mark and Diane Allen, Helen Darabant, Ted and Betty Kowalski, Georgine Bowlby, Paris Jenkins, Jim Fite and Paul Safchuck of the White Lung Association, Leslie Schneider, Leone Gordon, Bob Howard, Scott Schneider, Arlette Klaric (co-curator of the *Breath Taken* exhibition at Boston University), Bob D'Attilio, Michael Thornton, David Ozonoff, Paul Brodeur, Dr. Irving J. Selikoff, Barry Castleman, Ron Motley, Macy Lawrence, Chris Placitella, Judge Andrew McQueen, Lawrence Madeksho, Robert Ballard, Jonathan Bennett, Dr. Max Borow, Merilyn Jackson, Dr. Karen Antman, Dr. Russell Harley, Arthur Sabatini, Dr. Christine Oliver, Kaye Pyle, Martha Montgomery, Jerry Liebling, Jean Caslin, Stan Trecker, Brenda Cole, Tom Callahan, Marguerite Cavalero, Cathy Rodgers, David Kotelchuck, Craig Easson, Sheron Rupp, Carol Schloss, Mary Ann Hnatiw, Robert Bamford, Marie Bustard, Dr. Daniel Goldstein, Michael Moore, Leo Santa Maria, Francis Boudrow, Buddy McCourt, Alan Silverman, Harris Barron, Debbie Socolar and Dan Spikol.

Additionally, I would like to acknowledge the following organizations for their assistance and support: Asbestos Victims Special Fund Trust, White Lung Association, Association to Improve Benefits for Retired Disabled Workers and Their Dependents, Boston University Art Gallery, Boston University School of Public Health, Health/PAC, Center for Environmental Management/Tufts University, Drexel University, Photographic Resource Center, Houston Center for Photography, Manville Public Library, Massachusetts Cultural Alliance, Focus Infinity Fund, Ness, Motley, Loadholt, Richardson and Poole Law Offices, Thornton and Early Law Offices, Law Offices of Diane Middleton, the Lyda Ebert Family Foundation, the George Arents Research Library, and Local 6 Asbestos Workers Union.

This catalogue is dedicated to
the memory of my father, Anthony Ravanesi,
who died from mesothelioma in 1981.

Preface

Some years ago, commenting on his work about mercury poisoning in Minamata, Japan, the photojournalist W. Eugene Smith, stated, "To cause awareness is our only strength."

In the autumn of 1980, I was informed that my father had malignant mesothelioma. At that time, neither my father nor I had ever heard of this disease. We would soon learn that my father had only months to live, and that this fatal cancer was caused by his past exposure to asbestos as a shipyard worker in Boston during World War II.

Prior to his illness, my father was 6 feet tall, 220 pounds. In the two months before his death, his weight fell to 160 pounds, he was in constant pain, and was put on intravenous morphine. My family watched, stunned, as he wasted away. Finally, on January 9th, 1981, he succumbed to the cement-like tumors that had invaded the lining of his chest cavity.

A few years later, I came across *Expendable Americans,* written by Paul Brodeur. This book revealed to me an incredible story of how thousands and thousands of American men and women die each year of preventable diseases caused by exposure to asbestos. My own agonizing encounter with asbestos, together with my newfound awareness from Brodeur's book, led me to asbestos as the subject of my next project. *Breath Taken: The Landscape and Biography of Asbestos,* which began in January 1984, took me first to many research libraries, then to the homes of hundreds of victims and their families. As well, I visited the primary sites of the asbestos industry in the United States and Canada, many of the remaining shipyards, and spoke to professionals in the fields of science, medicine and law. I did all of this in an attempt to use my art to document this avoidable human disaster, and to come to grips with my outrage.

During the early phases of the asbestos project, while I was concentrating on oral history work, it became clear to me that the visual disposition of many of my subjects was very different from my preconceived notion. Of course, images of my father's suffering were still fresh in my mind's eye. I had expected that the victims I would meet would have had a similar fate. Although almost all of the victims had scarred lungs, and many were in different stages of various asbestos-caused cancers, many had an outward appearance of health. How does one photograph a victim, who for the most part, looks healthy, but is wasting away from the inside out? In the process of trying to deal with this reality I grew frustrated. Near the end of the first year of the project, I decided to select a small group of victims and to photograph and rephotograph them over a long period of time, creating sequences in which I also included their family snapshots, wedding pictures, and vintage images made earlier in their work careers by company photographers. Some of these sequences stretch over a 40-year period and document the progression of their diseases. In other cases, it became necessary to include their voices so that they could explain that despite their seemingly healthy appearance, they were indeed suffering. This led me to include oral history transcriptions in the exhibition narrative and to produce a video installation piece so that their testimonies could be heard.

Amid the controversies over liability for past exposures and prevention of future instances, the victims of asbestos-caused diseases were and still are strangely missing from our sight. To the degree that we see them at all, it is usually as objects rather than as subjects—statistics to be recorded, cases to be diagnosed, or plaintiffs to be deposed. Our knowledge of asbestos as a major medical, legal, and social problem has tended ironically to obscure the fact of asbestos as a profound human tragedy that many families have lived and continue to live through. *Breath Taken,* through its inclusion of both contemporary and vintage images, narrative, industry advertisements, objects, and voice will, I would hope, give the viewer a landscape of awareness of this human tragedy.

Today many people consider the asbestos problem behind us. The EPA has issued a phase-down ban, but what about the 30 million tons of asbestos fibers that remain in place in our society? Asbestos is in buildings like schools, homes, offices and the workplace; in automobiles; in the more than 200,000 miles of transite asbestos drinking water pipe. We, therefore, continue to be faced with important decisions regarding the public's health and safety. Will our public policies regarding this health menace add to the quarter million deaths already projected for the next 20 to 25 years from asbestos-induced lung cancers alone? Will all this breath taken be in vain?

Bill Ravanesi
Boston, Massachusetts

UNIVERSAL SAFETY
DO YOU BELIEVE IN SAFETY?
UNIVERSAL SAFETY
FIRE STATION Nº I
XI-837 ELEC

The Asbestos Tragedy

by Paul Brodeur

Paul Brodeur is a staff writer for The New Yorker, *and a leading investigative journalist on asbestos issues His books,* Expendable Americans *and* Outrageous Misconduct, *have chronicled the history of the asbestos catastrophe. Most recently he has been widely acclaimed for his 1989 book on electromagnetic fields,* Currents of Death.

Back in the nineteen-twenties, when the word "ASBESTOS" frequently appeared in large block letters on the fireproof drop curtains that closed off the proscenium openings of theatre stages, it was customary for youngsters attending Saturday-matinee performances of "Charley's Aunt" or "The Vagabond King" or "Mrs. Wiggs of the Cabbage Patch" to greet the raising of the asbestos curtain with the chant "All school boys eat stewed tomatoes on Saturday!" This was not, of course, an exercise in mnemonics but an expression of youthful exuberance, and, to a lesser extent, perhaps, a reflection of the tendency in an era celebrated for parochialism and flippancy to dismiss strange and exotic substances with jokes. Among older people, it was certainly common knowledge that asbestos was non-combustible, for when "Joe and Asbestos"—the syndicated comic strip about horse racing—first appeared in the Baltimore Sun, in 1925, the stableboy named Asbestos explained to the readership that he had acquired his odd moniker because "no matter how hot my tips are, they'll never burn a hole in your pocketbook." Such knowledge was hardly universal, however, for it was also true that a popular put-on of the flapper age consisted of solemnly assuring a susceptible theatre companion that the funny-looking word on the curtain meant "Welcome" in Latin. The fact of the matter is that although asbestos was linked one way or another in the public mind with fireproof theatre curtains, many people thought its name was a trademark, few people knew that it was being utilized in a wide variety of industrial products, only a handful of medical researchers were just then beginning to suspect that it was affecting the health of the workers who handled it, and virtually no one had any idea that it would one day prove to be one of the most important industrial causes of cancer in the world.

Asbestos, which comes from a Greek word meaning inextinguishable or unquenchable, is a broad term embracing a number of fibrous silicate minerals that are found in practically every country in the world. Although known to mankind as early as the Stone Age, when it was used in pottery, asbestos was considered a novelty until the late 1870s, when it began to be mined on a commercial basis. It soon achieved enormous industrial importance because of its unique and astonishing physical properties.

Asbestos appears to be highly combustible, yet it can withstand the fiercest heat. It seems as perishable as grass, but by virtue of being almost immune to the forces of corrosion and decay under almost every condition of temperature and moisture (and of being resistant as well to the action of most acids, alkalies, and other chemicals) it is just about indestructible. It looks extremely fragile, yet its fibers have a tensile strength equal to that of piano wire. Apparently as light and feathery as thistle or eiderdown, it is actually as heavy and dense as the rock from which it is extracted. In one sense, then, it is a fiber of stone. In another sense, however, it is a mineralogical vegetable whose fibers are so soft and flexible that they can be carded, spun, and woven as easily as fibers of cotton or flax.

Asbestos is the only mineral that can be woven into cloth, and its fibrous structure is, if anything, even more amazing than its remarkable ability to withstand heat. In fact, if it were not for the electron microscope, the extent to which asbestos is fibrous would be difficult to believe, for there are approximately a million individual fibrils lying side by side in a linear inch of chrysotile asbestos—a variety of the mineral that accounts for 95% of the world's production—whereas only 3,800 glass fibrils, such as those found in various insulation materials, or 630 human hairs can be aligned along the same distance. Moreover, in addition to their extreme fineness, high tensile strength, and unusual flexibility, spinnability, and resistance to heat and the elements, asbestos fibers possess great powers to adsorb and to filter. Small wonder, then, that when asbestos, which had been known to the ancients as "the magic mineral," was, in effect, rediscovered a little more than a century ago in the age of industrial expansion, it was put to work.

During this century, the demand for asbestos rose almost as fast as the stuff could be mined from the earth and milled from its host rock. In 1879, when the world's first commercial asbestos mine was opened at Thetford, in the Province of Quebec, only 300 tons of the mineral was produced. By 1910, annual world production had jumped to 30,000 tons; by the middle thirties, it had increased to 500,000 tons; and by 1970 the total output was more than four million tons a year. For several decades, asbestos was called "the mineral of a thousand uses" by those who sold it, but by the late 1960s this label had become obsolete, especially in the United States, where it is estimated that asbestos may have as many as 3,000 different industrial applications. As a result, asbestos has become practically ubiquitous in modern society. There is not an automobile, airplane, train, ship, missile, or engine of any sort that does not contain asbestos in some form or another, and it has found its way into almost every building, factory, home, and farm across the land. And, because its minuscule fibers are eminently respirable, asbestos has also found its way into the lungs of human beings, where, by remaining as indestructible as it does in nature, it can wreak terrible havoc.

The adverse biological effects of asbestos were observed as early as the first century by the Greek geographer Strabo and by the Roman naturalist Pliny the Elder, both of whom mentioned in passing a sickness in the lungs of slaves whose occupation was to weave asbestos into cloth. Strabo and Pliny were calling attention for the first time in history to a disease that would one day be known as asbestosis—a form of pneumoconiosis (the general term for all dust diseases of the lung) caused by the inhalation of the fine fibers and particles of asbestos. The use of asbestos in Europe appears to have diminished greatly during the Medieval period, but with the advent of the industrial revolution, asbestos came into widespread use. As a result, the fact that the magic mineral could produce lung disease, which had been forgotten since Strabo and Pliny first recorded it around the time of Christ, soon manifested itself again.

Modern knowledge of asbestosis dates from the year 1900, when a post-mortem examination was performed by Dr. H. Montague Murray, a physician in London's Charing Cross Hospital, on the body of a 33-year-old man who had worked for 14 years in an asbestos-textile factory. The patient, found to have been suffering from severe pulmonary fibrosis, which is scarring of the lungs, had been the last survivor of a group of ten men who were working in the carding room of the factory in 1886, and since Dr. Murray found spicules of asbestos in the lung tissues at autopsy, he was able to establish a presumptive connection between the man's occupation and the disease that killed him.

It was not until 1924 that the first clear case of death due to asbestosis appeared in medical literature. That year, Dr. W. E. Cooke, an English physician, who gave the disease its name, performed a post-mortem examination on a 33-year-old woman patient who had started working at the age of thirteen in an asbestos-textile factory. By 1917, after thirteen years of exposure, she had been coughing and in bad health. The autopsy showed extensive lung scarring and dense strands of abnormal fibrous tissue connecting the lungs and the pleural membranes surrounding them.

Cooke's discovery, which was published in the British Medical Journal, was the point of departure for an intensive study of asbestosis in Britain over the next seven years. Here in the United States, no investigations of the health experience of asbestos workers were undertaken by the United States Public Health Service. As early as 1918, however, the U. S. Bureau of Labor Statistics published a report by Frederick L. Hoffmann, a consulting statistician for the Prudential Insurance Company of America, who not only called attention to the fact that American asbestos workers were experiencing unusually early deaths, but also revealed that it had become the practice of some American and Canadian insurance companies not to issue life-insurance policies to asbestos workers, because of the "assumed health-injurious conditions" that existed in the asbestos industry.

By the early 1930s, asbestos workers who had developed asbestosis were bringing damage suits against Johns-Manville, of New York City, the largest asbestos manufacturer in the nation, and against Raybestos-Manhattan, of Bridgeport, Connecticut, the second largest asbestos company. As a result, the two firms, together with other leading asbestos manufacturers, initiated a cover-up of the asbestos hazard that continued for more than forty years. Here are some of the highlights of that cover-up:

In 1933, Lewis Herold Brown, the president of Johns-Manville, advised the company's board of directors that eleven pending lawsuits brought by employees who had developed asbestosis while working at the company's plant in Manville, New Jersey, could be settled out of court, provided that the attorney for the plaintiffs could be persuaded not to bring any more cases against the company.

In 1934, Vandiver Brown, the head of Johns-Manville's legal department, persuaded Dr. Anthony J. Lanza, a physician at the Metropolitan Life Insurance Company, to delete unfavorable information from a report about disease among asbestos workers that was soon to be published by the U.S. Public Health Service.

In 1935, Sumner Simpson, the president of Raybestos-Manhattan, wrote Vandiver Brown a letter, telling him that "I think the less said about asbestos the better off we are," to which Brown replied, "I quite agree with you that our interests are best served by having asbestosis receive the minimum of publicity."

In 1936, Vandiver Brown and Sumner Simpson, together with officials of several other asbestos manufacturing companies, arranged to finance animal experiments at the Trudeau Foundation's Saranac Laboratory, at Saranac Lake, New York, in order to gather data that they expected would help the asbestos industry defend against claims brought by workers who had developed asbestosis. The studies showed, however, that significant numbers of animals developed asbestosis after being allowed to inhale asbestos. These results were suppressed by the asbestos manufacturers for more than forty years.

In 1943, Vandiver Brown told representatives of the Union Asbestos & Rubber Company, of Paterson, New Jersey, that Johns-Manville did not inform its employees when their chest X-rays showed that they had developed asbestos disease. Brown said that if the workers were told, they would stop working or file claims against Johns-Manville, and that it was company policy to let them work until they quit because of asbestosis, or died as a result of it.

In 1949, Dr. Kenneth W. Smith, who would subsequently become the medical director of Johns-Manville, sent a memorandum to company headquarters concerning seven asbestos-mill employees whose X-rays showed signs of early asbestosis. The memo, which did not come to light until 1976, provided written evidence that Johns-Manville was following a policy of not informing its workers when they developed the disease. "The fibrosis of this disease is irreversible and permanent so that eventually compensation will be paid to each of these men," Smith wrote. "But as long as the man is not disabled it is felt that he should not be told of his condition so that he can live and work in peace and the company can benefit by his many years of experience."

In 1952, the Seventh Saranac Symposium—a week-long meeting on pulmonary dust disease—was held at the Saranac Laboratory. It was attended by more than two hundred medical doctors, research scientists, state and federal public health officials, insurance-company executives, and asbestos manufacturers, who were told about medical evidence implicating asbestos as a potent lung-cancer producing agent. Unlike the proceedings of six previous symposia, the proceedings of this meeting were never published. And because only one or two of the participants spoke out about what they had heard, almost no infomation about the carcinogenicity of asbestos found its way into the press for another decade.

The result of the pervasive cover-up of the asbestos-disease hazard was a national public health disaster of unparalleled magnitude, which is unfolding to this very day. Some twenty million unsuspecting American workers—four and a half million men and women in the wartime shipyards alone—underwent exposure to dangerously high levels of asbestos dust as they applied, or worked near people who were applying, asbestos insulation to boilers and high-temperature pipes in ships, power plants, oil refineries, and chemical factories; as they sprayed asbestos insulation on the steel girders of high-rise buildings, or worked at trades in close proximity to such operations; and as they toiled in mines from which raw asbestos was extracted, and in factories where asbestos products were manufactured. Hundreds of thousands of these people have either developed or will develop incurable asbestos disease. Tens of thousands of them have died of it.

The asbestos cover-up might have gone on indefinitely had it not been for two extraordinary developments during the early and middle 1960s—one in law and the other in medicine—which would ultimately result in exposing the misconduct of the asbestos manufacturers and making them accountable to some of their many victims. The legal development occurred in 1965, when the American Law Foundation redefined tort law to make the sellers of all unreasonably dangerous products strictly liable to users and consumers unless their products carried adequate warning labels. The medical development consisted of some pioneering epidemiological studies of the health and mortality experience of asbestos-insulation workers which had been conducted during 1962 and 1963 by Dr. Irving J. Selikoff, director of the Mount Sinai School of Medicine's Environmental Sciences Laboratory, in New York City; Dr. Jacob Churg, chief pathologist at Barnert Memorial Hospital, in Paterson, New Jersey; and Dr. E. Cuyler Hammond, vice-president for epidemiology and statistics of the American Cancer Society.

Dr. Selikoff gave X-ray examinations, pulmonary-function tests, and blood tests to 1,117 asbestos-insulation workers who were members of New York Local 12 and Newark Local 32 of the International Association of Heat and Frost Insulators and Asbestos Workers. He found radiological evidence of asbestosis in fully half of these men. Moreover, he found that among 392 men with more than twenty years of exposure, 339 (slightly less than 87%) had developed the disease. Even more alarming were the results of a carefully conducted mortality study of 632 workers who were on the union rolls at the two locals as of December 31, 1942. According to the standard mortality tables, 203 deaths could have been expected among the 632 workers. Instead, there were 255, not counting seven men who had died before incurring twenty years of exposure—an excess of 25%. The reason for the excess was not hard to find: these men had succumbed to lung cancer at seven times the expected rate, and to gastrointestinal cancer at three times the expected rate.

The studies of Dr. Selikoff and his associates furnished incontrovertible evidence that industrial exposure to asbestos was extremely hazardous, and they marked a turning point in the views held by doctors and health officials in many parts of the world.

Unfortunately, state and federal health officials in the United States were lamentably slow in reacting to the studies of Selikoff and his colleagues. Indeed, in some cases they were clearly reluctant to take any action at all. During the 1960s, the chief industrial hygienist for the U.S. Public Health Service's Division of Occupational Health entered into confidentiality agreements with asbestos manufacturers that prevented him from giving out any details concerning the asbestos exposure of any of the workers employed at the asbestos factories his division was inspecting. As a result, the Public Health Service did not make any recommendations to asbestos workers or to their unions about how workers might protect themselves from the hazard of toiling in excessive dust levels. For its part, the U.S. Department of Labor merely saw fit to reduce its ludicrously inadequate and almost totally unenforced standard for occupational exposure to asbestos from one that allowed workers to inhale tens of billions of asbestos fibers and fibrils each day, to one that allowed them to inhale hundreds of millions of fibers and fibrils each day. As for the Environmental Protection Agency and its predecessors, these organizations allowed asbestos insulation to be sprayed on the steel girders of high-rise buildings until 1972, when this extremely hazardous practice was banned nationwide, and to be used in the construction of thousands of schools, offices, and apartment buildings across the United States.

By this time, asbestos-insulation workers were bringing product-liability lawsuits against the manufacturers of asbestos insulation, who had failed to attach labels to their products warning that asbestos could cause disease. As might be expected, the studies of Dr. Selikoff and his associates were of considerable importance in helping these plaintiffs establish the health hazard of asbestos. Equally important was the determination of Ward Stephenson, a trial lawyer from Orange, Texas, who won the first asbestos product-liability lawsuit, in 1971. This landmark verdict was upheld by a federal court of appeals in 1973, and it paved the way for the greatest toxic tort litigation in the history of American jurisprudence.

During the next ten years, Johns-Manville, Raybestos-Manhattan, and more than a dozen other manufacturers of asbestos insulation were the targets of some 15,000 lawsuits. At first, the defendant manufacturers tried to claim that they did not know about the asbestos hazard until Dr. Selikoff's pioneering studies of the early 1960s. However, plaintiff attorneys soon unearthed Sumner Simpson's correspondence, Vandiver Brown's letter asking Metropolitan Life to water down its report about asbestos disease, Dr. Smith's memorandum about why diseased asbestos workers were not informed of their condition, and hundreds upon hundreds of other documents that furnished overwhelming proof that officials of Johns-Manville, Raybestos-Manhattan, and other leading asbestos companies had known about and engaged in a cover-up of the disease hazard of their products for more than forty years. As a result, juries from one end of the nation to the other began awarding large compensatory damages to diseased asbestos workers and the survivors of workers who had died of asbestos disease. They also began assessing punitive damages against asbestos manufacturers for outrageous and reckless misconduct. Indeed, during 1981 and the first half of 1982, juries in ten different cases found Manville liable for punitive damages totalling more than six million dollars. The company and its insurance carriers had already settled some 2,000 asbestos-disease cases out of court for tens of millions of dollars.

A new era in the asbestos tragedy began in August of 1982, when Manville—a corporation with assets of more than $2 billion—filed for protection under Chapter 11 of the federal Bankruptcy Code, claiming that it had been unfairly victimized by thousands of unwarranted lawsuits. During the next six and a half years, while the affair languished in the Dickensian coils of the federal bankruptcy system, Manville was not required to pay a single dime to any of its thousands of victims. Moreover, when the final bankruptcy plan was approved, it made provision for compensating only 100,000 asbestos-disease victims.

This estimate was absurdly low by any reasonable standard of judgment. To begin with, Dr. Selikoff and his associates have estimated that between 200,000 and 300,000 people will die of asbestos-related lung cancer during the next twenty to thirty years. Secondly, of the 18 million or so American men and women still alive who have undergone occupational exposure to asbestos, it can reasonably be expected that as many as two million of them may develop X-ray evidence of asbestos disease. Third, studies already conducted indicate that fully 30% of the wives, children and family members of heavily exposed asbestos shipyard and factory workers—some 200,000 people in all—are developing X-ray evidence of asbestos lung disease. Fourth, other studies show that about 30% of the school custodians in Boston, New York, and San Francisco are developing X-ray evidence of asbestos disease. And, finally, surveys presently underway indicate that about 30% of the nation's merchant seamen, who have been exposed to asbestos by virtue of having worked in the confines of ships, are also developing signs of asbestos disease. Thus, to assume that only 100,000 of nearly three million potential asbestos-disease claimants would bring claims was clearly preposterous. Indeed, there are now 150,000 asbestos-disease victims who have filed claims against the Manville Trust and other asbestos manufacturers.

From the foregoing, it seems clear that the extent of the asbestos tragedy has yet to be ascertained, and that when the statisticians have tallied the final toll, the casualties will have been far more numerous than our present society and its legal system is willing to acknowledge. Statistics can, of course, be a way of dehumanizing and thus denying the true dimensions of the misery that afflicts asbestos victims and the victims of other industrial and environmental disease who live among us. Fortunately, as an antidote to this possibility, we now have "Breath Taken," Bill Ravanesi's moving chronicle of the asbestos tragedy. By focussing upon the human condition, Ravanesi's splendid photographs demonstrate the inadequacy of statistics. Indeed, by revealing the pain, bewilderment, despair, and resignation etched in the faces of asbestos victims, he provides us with an indelible reminder that statistics are human beings, with the tears wiped off.

Taking Photographs, Making Statements: Bill Ravanesi's Asbestos Project

by Arthur J. Sabatini

Arthur J. Sabatini is a writer and teacher. He specializes in the areas of contemporary art and performance, critical theory, and cultural studies. He holds a Ph.D. from New York University and has taught at New York University, University of the Arts, and Drexel University.

As a title for an exhibition, *Breath Taken: The Landscape and Biography of Asbestos* sounds ambiguous and threatening.

Does 'breath taken' refer to breathing done in the past, or a single breath once taken? Or, does it connote having one's breath 'taken' away? What could that mean? Taken by whom? Why? Or, could the phrase allude to the exclamation 'breathtaking'? Breathtaking is commonly used to express awe at the sight of a panoramic natural landscape—a canyon at sunset, snowcapped mountains, or a beautiful artistic reproduction of grand vistas. But landscapes are not always natural, they can be man-made. Does the play on words here suggest that there is something about the 'landscape' that has taken breath away? In this context, moreover, the word 'landscape' is coupled with 'and biography of asbestos.' This seems even more troubling and unnatural. Asbestos, a mineral that has become synonymous with numerous manufactured products, is a natural fiber embedded in serpentine stone in landscapes around the world. It is discernable, as several of Bill Ravanesi's detailed photographs reveal, in whitened streaks and clusters in jagged, dark greenish rock formations. Asbestos, then, is a raw feature of certain landscapes, yet it is also 'landscaped' in the process of earth brutalizing strip mining in open pit quarries. But, even knowing this, in what ways can asbestos be said to have a biography? Is Ravanesi implying that organic life can be attributed to this inert substance? Or, that, somehow, human biographies are connected with asbestos? If so, what is the relationship between asbestos and breath? And, as if these questions are not complicating enough, exactly what type of exhibition—artistic, scientific, journalistic, medical—is promised by such an ominous and, it should be noted, carefully chosen title?

These questions, and others, are quickly answered as you become familiar with the intentions and range of Bill Ravanesi's project. The following commentary will describe the *Breath Taken* exhibition, and discuss how he works in the field as a photographer, researcher, and organizer. For *Breath Taken* is a coordinated display of photographs, objects, written documentation, and events that focus on asbestos and its deleterious impact on human lives. After experiencing *Breath Taken,* which is often accompanied by conferences and symposia, you find yourself asking far deeper and more unsettling questions about the issue of asbestos, the landscapes where it is found, and the biographies it has affected. It then becomes apparent that with *Breath Taken,* Bill Ravanesi is not involved in word play nor the pursuit of artistic expression. The knowledge he has accumulated and the people he has lived with and photographed compel him to make very unambiguous statements about this mineral and the abuses of the asbestos industry.

Bill Ravanesi is a photographer and committed activist. He wants the images and supplementary materials in *Breath Taken* to stun you into a visceral realization of the depth and horror of what Paul Brodeur calls "the tragedy of asbestos." As a result of proven maleficence of forces in the manufacturing and insurance industries, and government, concerning the dangers of asbestos, hundreds of thousands of people have died and many are at risk from the 30 million tons of asbestos still in place in communities and the workplace.

There are several torturous and fatal varieties of asbestos-caused diseases, most of which affect the lungs. The individuals Ravanesi photographed for *Breath Taken*, many of whom have since died, stand weakly or sit with their bodies contorted as they struggle for breath. Their lungs are degenerating from scars or tumors. Their fingers, like those of Clementine Szukis and Tom Bowlby, are clubbed because of their lungs' inability to move oxygen to their extremities. [see pages 29 and 34] Eyes, like those of former boilermaker Paris Jenkins, are swollen and rheumy because of microscopic asbestos fibers embedded under the lids. [see page 19] Collectively, their expressions tell you that they know they did not deserve to be suffering from debilitating illnesses. That, if they had been informed of the risks of working with asbestos, or about the hazards of living near asbestos factories, they could have avoided the pain you see. Ravanesi's confrontational and deeply felt portraits, unlike typical 'social issue' documentary photographs, do not particularly invoke sympathetic responses. Instead, by setting up these images with cool lighting and black drapes, he demands that you study the faces and reflect on the lives of these victims and their condition.

As a contemporary visual artist, Ravanesi pursues what I call "aesthetic research." He is working toward a recognition of a socially grounded practice that expands the boundaries of documentary photography, art, investigative journalism, and social scientific research. Throughout his career, Ravanesi has undertaken similar, long-term projects. But *Breath Taken* is more extensive than his work covering migrant farm workers and his efforts with urban minorities in Holyoke, Massachusetts and other cities.

Shingle Punch Crew, Johns-Manville Plant, Manville, NJ 1948

With *Breath Taken* Ravanesi is, in effect, attempting to photograph something more insidious, widespread, and virtually invisible. Moreover, in *Breath Taken* he aggressively identifies the responsible parties. Thus, Ravanesi is arguing that conventional or "objective" types of inquiry and existing artistic or journalistic forms of representation that "speak out" about health and the environment do not adequately convey the depth or complexity of "the asbestos issue." In order to understand asbestos—as a mineral, globally utilized industrial product, and cause of disease—Ravanesi has felt it necessary to combine the methods and tools of art, visual technology, reportage, and social activism. Throughout the seven year project, he lived in asbestos mining and manufacturing towns. He photographed, listened to, and recorded the stories of hundreds of asbestos workers and victims. He confronted officials from the industry and provoked interaction among physicians, lawyers, union officials, workers and policy makers.

In many ways, Ravanesi builds on the photographic legacies of Jacob Riis, Lewis W. Hine, Eugene Smith, and the Depression era Farm Security Administration; but he is equally indebted to the probing life history studies of Robert Coles, Hans Haake's politicized art exhibitions, the gritty documentaries of filmmaker Frederick Wiseman, and the exemplary investigative journalism of Paul Brodeur. Ravanesi's concerns are overtly social and artistic, although his subject matter is, obviously, the stuff of business, physical science and medicine.

But, why, he asks, should we think in such exclusive categories? Why shouldn't artists explore subjects like asbestos and seek ways of representing complex social, industrial, or medical themes, especially when human lives and the environment are at stake?

Although the content of each exhibition varies, in its most complete version *Breath Taken: The Landscape and Biography of Asbestos* includes up to 90 large format color photographs and a 36 minute video of testimony by victims of asbestos-caused diseases and their leading medical, environmental, and journalistic advocates. These are integrated with 23 actual asbestos products; asbestos industry advertisements, packaging, and archival documents relating to legal and medical aspects of workers' conditions and health risks; 35 vintage photographs by Margaret Bourke-White and others on the triumphs of the asbestos manufacturing process; a 2,000 item bibliography of scientific reports and other articles on the asbestos crisis; and what Ravanesi calls a "Wall of Shame," which lists the names of asbestos corporations, insurance companies, members of the medical-industrial complex, and those who have been responsible for policies that have adversely impacted on workers and victims. Every image and object in the exhibition is captioned and a factual running commentary (in wall texts) by Paul Brodeur charts the machinations of the industry's conspiracy to deny the effects of asbestos fibers on workers' health. When displayed, the 175 piece presentation wraps around walls, occupies enclosed rooms, fills showcases, and tops pedestals.

As you take in the *Breath Taken* exhibition you realize that asbestos has a long, tangled history. First discovered in ancient Egypt and Rome, asbestos is enmeshed in our lives materially, economically, and ecologically. It is naturally abundant in regions worldwide and a component of thousands of products used in industrialized societies. Samples of raw asbestos and items such as house shingles, a toaster, automobile brake pads, insulation, and fire protective gloves are placed throughout the exhibition. Some are sealed in plexiglass since one asbestos fiber inhaled can lodge in the chest for decades and slice through the tissue of the lungs like a razor. Ravanesi has photographed eight gross specimens of scarred, fibroid, tumorous lungs, and chest cavities decimated by asbestos dust. One image coldly titled "Mesothelioma Gross Specimen" shows a pair of lungs and a trachea in a glass case. The left lung is crusted in a coral colored mass, as if in the grasp of a mutant conch shell. Like the portraits of victims, the lungs are isolated in front of a symbolic black drape.

The asbestos industry's promotional materials, vintage photographs, and Margaret Bourke-White's 1934 black and white photojournalism study for *Fortune Magazine* portray asbestos as an indispensable modern product. Male and female factory workers are shown weaving textiles or mixing asbestos cement in enormous, polished machines. Although supposedly dedicated, rugged, and proud—symbols of 'America at Work'—the workers seem posed, and indifferent, if not quietly resistant to being observed.

By contrast, Ravanesi's finely composed color photographs alternately thrust you into the bleak, ravaged mining sites and industrial landscapes where asbestos is gouged from the earth and processed. Related images pull you further into parking lots, factory grounds, and waste dumps. Eventually, Ravanesi leads you into the streets of the dreary company towns where most workers live. Portraits of the victims and photographic biographies of them and their families complete the exhibition along with photographs of individuals like Dr. Irving J. Selikoff and union officer Tony Mazzocchi, who have been leading the fight against the spread of asbestos-caused disease. (There are over forty portraits of the living and the dead in *Breath Taken*.)

Ravanesi's photographs are interrelated by content, thematically, and compositionally in various groupings. One way to view *Breath Taken* is along the lines of integrated series, unfolding sequences, or in varying thematic or conceptual clusters. In addition, the black metal-framed photographs serve as prismatic reflections, visual re-searchings, and aesthetic counterpoints to the written information, objects, and archival documentation in *Breath Taken*. For example, read narratively (as in the above paragraph), the mining and manufacturing of asbestos amounts to a 'biography' that functions as background to the human biographies. The objects in the exhibition tactilely affirm this. But, it is as consistent to see artistically formal or tonal analogies between, say, the architecture of geographically separate mine sites, factories, or cityscapes.

Ravanesi's photographs refer to each other in terms of settings, coloring, texture, and scale. Taken with a 4" x 5" Deardorff view camera, his images are obviously preconceived, posed, and acutely balanced. He develops each print himself and purposely drains the color from the landscapes while accenting minute physical details. There are no blue skies in any of Ravanesi's outdoor photographs; none of the images are hurried. Pictures taken in Vermont, Illinois, South Carolina, and Quebec reveal similarly scrubby terrains. [see page 25] In the portraits, he deepens and smooths backgrounds, sharpens the contrasts in facial color, heightens the rhythms of features, and, in several cases, produces larger than life-size murals.

About one-third of the images in *Breath Taken* are of mines and manufacturing plants, and the towns that surround them. In one study, reminiscent of those by 19th century landscape photographer Timothy O'Sullivan, Ravanesi sets his camera above a breathtaking panorama of a sloping valley in Lowell, Vermont, its hillocks and ravines slashed by the open pit VAG Mine. Another image holds you captive on a late afternoon before the awesome sprawl of the Johns-Manville Plant in Waukegan, Illinois.

Ravanesi's stately study of the Jeffrey Mine in Asbestos, Quebec telescopes the imagery of many other photographs. [see page 36] Measuring about a square mile and almost 1,200 feet deep, the Jeffrey Mine is the largest chrysotile mine in the Western hemisphere. Ravanesi's mesmerizing ashen grey, green, and dirt brown photographic dissection of this vast site is a summary statement on the beauty and might of industrial power as well as a commentary on the authority of landscape photography. The sizable 30" x 40" mural is detached and meticulous in its textures and subtly changing hues. The camera, perched on the upper ridge of the minescape, overlooks two milky pits at the bottom of a circular maze of roads that ribbon outward toward a mill situated at the top of a far ridge. Clumps of asbestos spill out from jagged rocks and the clawed body of the hills. Massive mustard colored tractors and heavy equipment, like randomly scattered children's toys, are pitched on the roads. An imposing white mill stretches across the top of the photograph. Constructed from transite, an asbestos-based material, the mill is an eleven story fortress of smokestacks, railings, and chutes. To its left, in a remote distance, trees are visible. A sulky oyster grey band of ragged clouds fill in the top border of the picture. In the far background, the town of Asbestos, Quebec sits precariously. At 3 PM each day it is rocked by explosions that loosen serpentine rock and asbestos dust.

A grim sense of purpose is exposed in the photograph of the Jeffrey Mine. Signs of force are everywhere; there are no people in the picture. Yet, for all that, this is a remarkable sight, as worthy of aesthetic representation as Babylonian ziggurats or the valleys and mountain ranges of America. Ravanesi is keenly aware of the fascination with capturing a sense of place and recording the "progress" of civilization. But, as we near the turn of this century, he implies, we can ill afford merely to take pleasure in aesthetic images without turning our attention to the realities of our relationship with each other and the environment.

As over a dozen of Ravanesi's photographs reveal, asbestos is at the center of the life of entire regions in the United States and Canada. Images of the Jeffrey Mine are accompanied by aerial photographs of the mine site and ten other views of the town of Asbestos, Quebec. A 30" x 40" mural titled "Asbestos Pavement and Cinema" starkly depicts a movie house on the corner of an empty street in downtown Asbestos. The faded sienna colored walls of the bunker-like building are coated with asbestos cement; the roof is asbestos shingling. The streets that frame the movie house seem laminated with a blue-black sheen. Like bits of shrapnel, flecks of asbestos are visible in the pavement. [see page 33]

A similar series taken in Manville, New Jersey contains an aerial view showing the proximity of the plant to the town's main street. Another image places you in front of the washed out exterior of the Manville Borough Municipal Building. Its dingy facade is made with asbestos sheeting donated by the Johns-Manville Corporation. The sense of an oppressed city is reinforced in a photograph titled "Boys Playing Caps." [see page 38] At first glance, this is a snapshot of a placid suburban street on a gloomy afternoon. Three children are playing with caps near the sidewalk in the foreground of the 20" x 24" print. The garages of two smallish, white houses, separated by a patchy green, brown lawn are centered on both sides of the photograph. An enormous Manville Borough water tower looms in the background, like a sinister steel blimp propped up by girders. The sight lines of the photograph formally pull you into the vanishing point at the center of the water tower. Water in Manville, a caption notes, flows through transite pipe constructed from asbestos and cement. Over 200,000 miles of similar gradually deteriorating piping are still in use in communities throughout America.

The Manville series also includes a photograph titled "Loading Dock No. 5 with Asbestos Dust," taken at the Johns-Manville Plant in 1986. [see page 20] This is the same factory featured in Margaret Bourke-White's 1934 article. But Ravanesi shows worn red brick, crumbling cement, and asbestos sheeted window panes surrounding huge, closed barnlike wooden double doors. The doors are rotting, their bluish paint peeling. The day is done; no one is loading anything at this dock. A weighted, billowing rubberized black curtain dusted with asbestos shrouds the clamped doors. The image is an allegory of the fate of the asbestos industry: ineptly sealed doors, cheap funeral ornamentation, a ruined factory so full of deadly contaminants that it can not be safely torn down. The ambient dust within, like the deposits on the loading dock doors, will eventually blow into the atmosphere around the community.

Those who work directly with asbestos or live near mines and plants are the most susceptible to disease. Since asbestos-caused diseases develop slowly, usually over decades, victims often look outwardly healthy while tumors fester, and the tissue in their lungs and chest decay. Long time asbestos-exposed workers manifest some observable physical abnormalities, such as finger clubbing and cyanosis. The most fatal forms of illness caused by asbestos are asbestosis, lung cancer, mesothelioma, and other cancers. [see Paul Brodeur's essay] In most cases, the first signs of illness are shortness of breath and fatigue. For some victims, like Joe Darabant, it might take fifteen years before the only way to breathe is through an oxygen tank. For Darabant, who was forced to retire at 50 and died at age 66 from asbestosis, it could be said that he had taken his last clean breath in 1941, on the day he began working in "E" Building at the Johns-Manville Plant in Manville, New Jersey. [see back cover]

Ravanesi relates Joe Darabant's biography through eight images that include re-photographed snapshots, mounted on asbestos shingles, of Darabant at work in the 1940s, on his wedding day in 1949, and at a retirement party in 1974. A late 1989 image, by Ravanesi, portrays Darabant with a sunken chest and whitening skin, gasping for air as he sits on the edge of his bed. [see page 21] He also poses before a black velvet cloth, in a wheelchair equipped with an oxygen tank. In yet another formal portrait of his upper body and face, Darabant wears a sheer white shirt, clear plastic tubes inserted into his nose cross his flushed cheeks. This 30" x 40" mural draws you into Darabant's painful situation, on a scale larger than in life. At the time, Darabant knew he was only months from death. Nevertheless, he managed to stand proudly before the camera. He stares with quiet anger, courage, and deep integrity at Ravanesi and whoever else will see the picture. [see cover]

The biographies of Tom Bowlby, and Ted and Betty Kowalski are similarly told. In addition to Ravanesi's prints, older and recent photographs from different sources are displayed in combination with medical records, mementos, and oral histories. Ravanesi calls this arrangement of images "sequential reveal." (The term is equally applicable to the landscape photographs.) His intention is to present the viewer with incremental and/or dramatic layerings of information, rather than singularly arresting prints. Thus, Ravanesi accentuates the fact that photography is a time bound, historically contingent medium that can responsibly and authoritatively serve socially conscious, humanistic ends. By re-photographing subjects, Ravanesi demonstrates (as have others) how sustained contact creates the conditions for visual and emotional revelations between photographers, their subjects, and audiences.

Quite obviously, Ravanesi could not have taken certain photographs or secured personal items from Joe Darabant or the others without having gained their trust. This is most evident in the hospital room and deathbed pictures of Darabant, Bowlby, and Father Richard Pankowski. [see page 26] Here, again, Ravanesi attempts to reformulate prevalent photographic practices. Unlike roving photojournalists, Ravanesi eschews the idea of seizing fleeting moments from perspectives that, implicitly, only the photographer can glimpse. Working with color emulsions and with a view camera mounted on a tripod meant carefully absorbing the sights around him, waiting patiently for cloudy days, and becoming involved with the lives of the people he photographed. (Occasionally, when unexpected or public events related to the asbestos issue occur, Ravanesi relies on his medium-format cameras and photojournalistic instincts. A photograph titled "Mrs. Pankowski, Manville, NJ 1986," depicts three widows and mothers consoling each other on the front steps of a small house after a funeral. [see page 27] Another shot holds an important moment in a demonstration by an occupational health group, PHILAPOSH, outside the City Hall Annex in Philadelphia.) [see page 23]

By choosing to take staged, posed studies of individual victims, families, workers, and advocates, Ravanesi is keyed into the formulas and history of portraiture, one of the most conventional forms of art and photography. Portraits, whether of family or business leaders, celebrities, or 'social types,' involve emotional complicity, business arrangements, or an artistic agreement between the photographer and his subjects.

For Ravanesi, portraits of victims immersed him in a more delicate ethical and photographic problem: generally unfamiliar and indifferent to art, his subjects were in pain and dying. They had little reason to be photographed or much to offer Bill except their trust, intimacy, and hope that somehow his work would help others. The results are portraits that, in addition to their formal strength and technical qualities, speak through the photographer and the camera directly to the viewer. 'Understand what has happened to me,' all the victims appear to be saying. 'And do something about it.' [see page 19]

Ultimately, Bill Ravanesi wants us all to do something about the companies responsible for asbestos-caused diseases. For, as those who have assessed the situation have proved, asbestos, which is still used and in place in structures throughout the world, is deadly. But, if we attend to its dangers we can, perhaps, learn to avert other health and environmental disasters. [see the essays by Brodeur, Kotelchuck, and Castleman] As a photographer and artist, researcher and activist, Ravanesi employs an array of resources to communicate these ideas. He is passionately committed to demonstrating that it is possible to see harsh, complex, and seemingly intractable social issues through aesthetic media, particularly photography. He has also brought into focus the lives of real people and communities that are rarely seen. The accomplishment of Bill Ravanesi's *Breath Taken* project can be witnessed in the photographs, in the materials he has assembled, and the ambitious steps he has taken toward visually articulating the threatening landscape and treacherous biography of asbestos.

Dum Spiro Spero

While I breathe, I hope

Paris Jenkins
Charleston, South Carolina, 1989

Mr. Jenkins is a former boilermaker at the Charleston Naval Shipyard. He is disabled with asbestosis. Charleston, South Carolina is the home of Raybestos-Manhattan (now called Raymark Industries, Inc.), a major asbestos textile manufacturing plant and also home of many shipyards. Today, the Charleston area has a significantly greater incidence of lung cancer than the rest of the United States.

The quote on the opposite page is from the state seal of South Carolina.

Loading Dock No. 5 with Asbestos Dust
Johns-Manville Plant, Manville, New Jersey, 1986

Finished asbestos products left this area for national and international distribution. According to industry reports, asbestos is in approximately 3,500 industrial and domestic products.

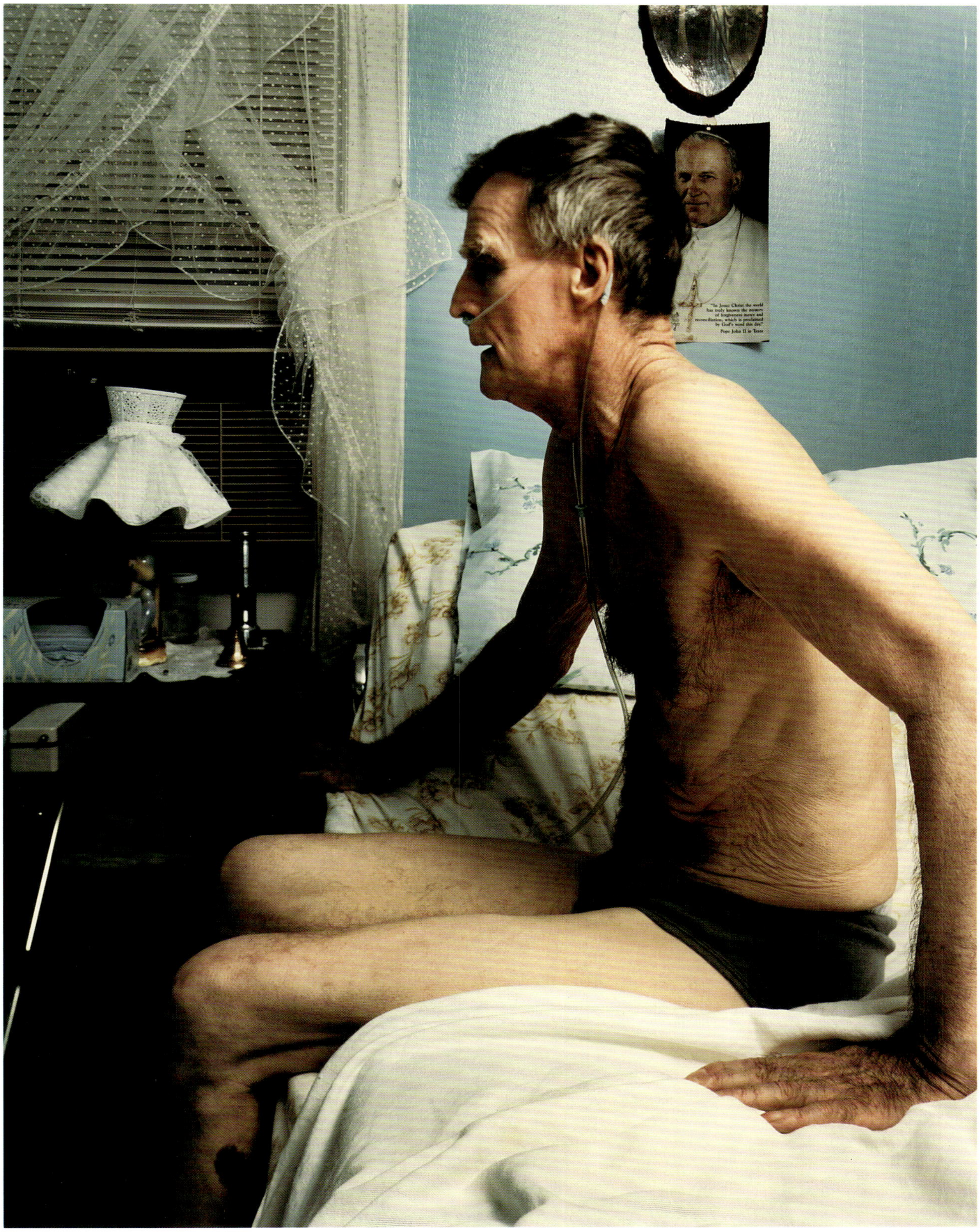

Joe Darabant
Manville, New Jersey, September 1989

This photograph was taken three months after the cover photograph. At this point, Joe was unable to feed himself and needed constant care. He died from asbestosis on April 26, 1990 at the age of 66.

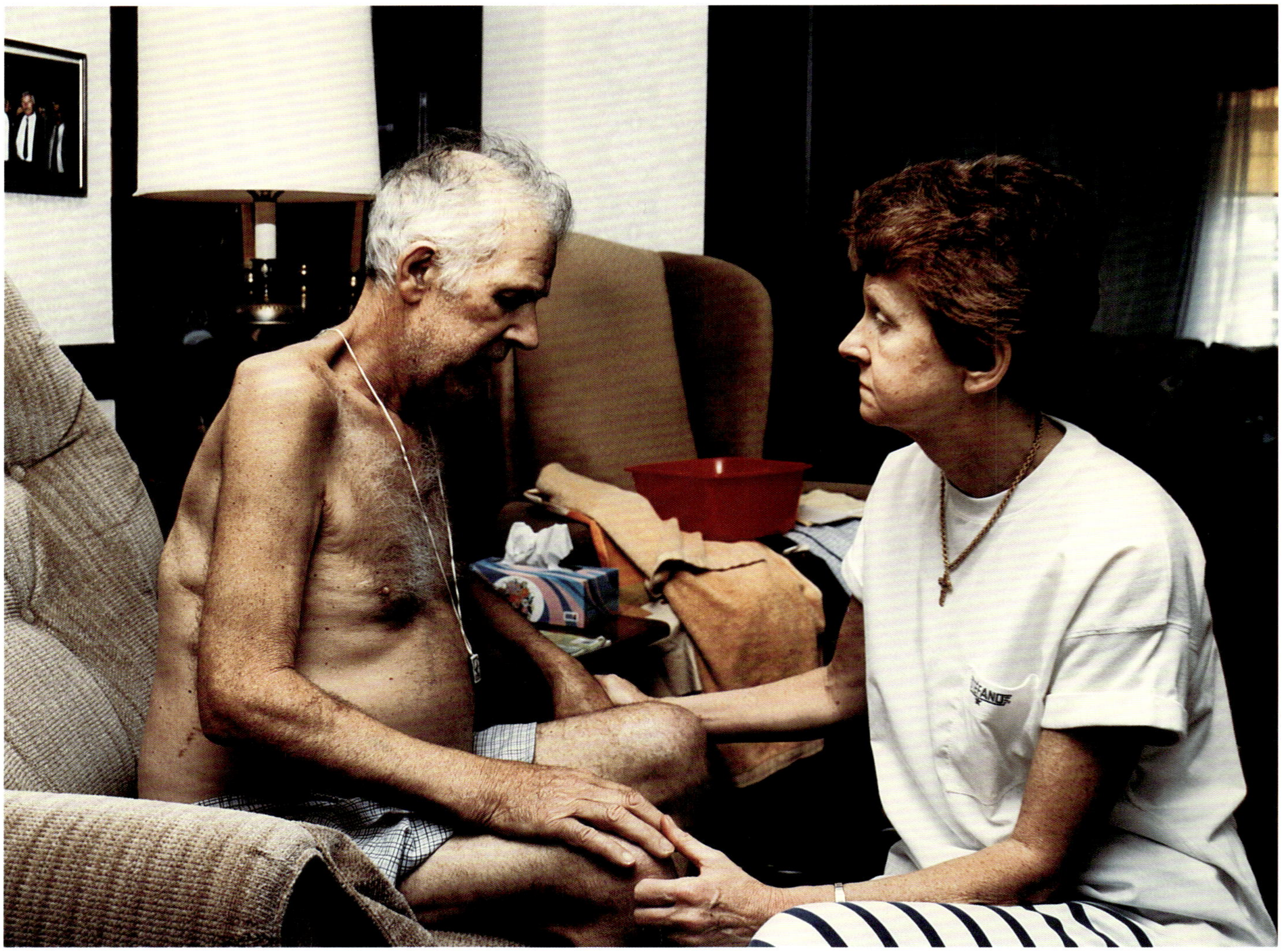

Jim and Mary Brienzo
Lynn, Massachusetts, June 1989

Former shipyard worker and fireman from Lynn, Massachusetts. Diagnosed with malignant mesothelioma early in 1989. He died the morning after this photograph was taken.

PHILAPOSH
Picketing City Hall Annex, Philadelphia, 1986

An occupational health group picketing to protest unacceptably high levels of ambient asbestos in this building.

Fred "Bud" Howard
Quincy, Massachusetts, 1989

Disabled with asbestosis, Mr. Howard died shortly after the portrait. He was a marine electrician at the Quincy Fore River Shipyard, Quincy, Massachusetts. He formed a local victims' organization called OFFER, the Organization of Federal and Former Federal Employees Rights.

Asbestos Tailings and Pond
Vermont Asbestos Group (VAG) Mine
Lowell, Vermont, 1988

This is one of the last operating chrysotile mines in the United States. Most of the asbestos mined at this site is sold to third world countries.

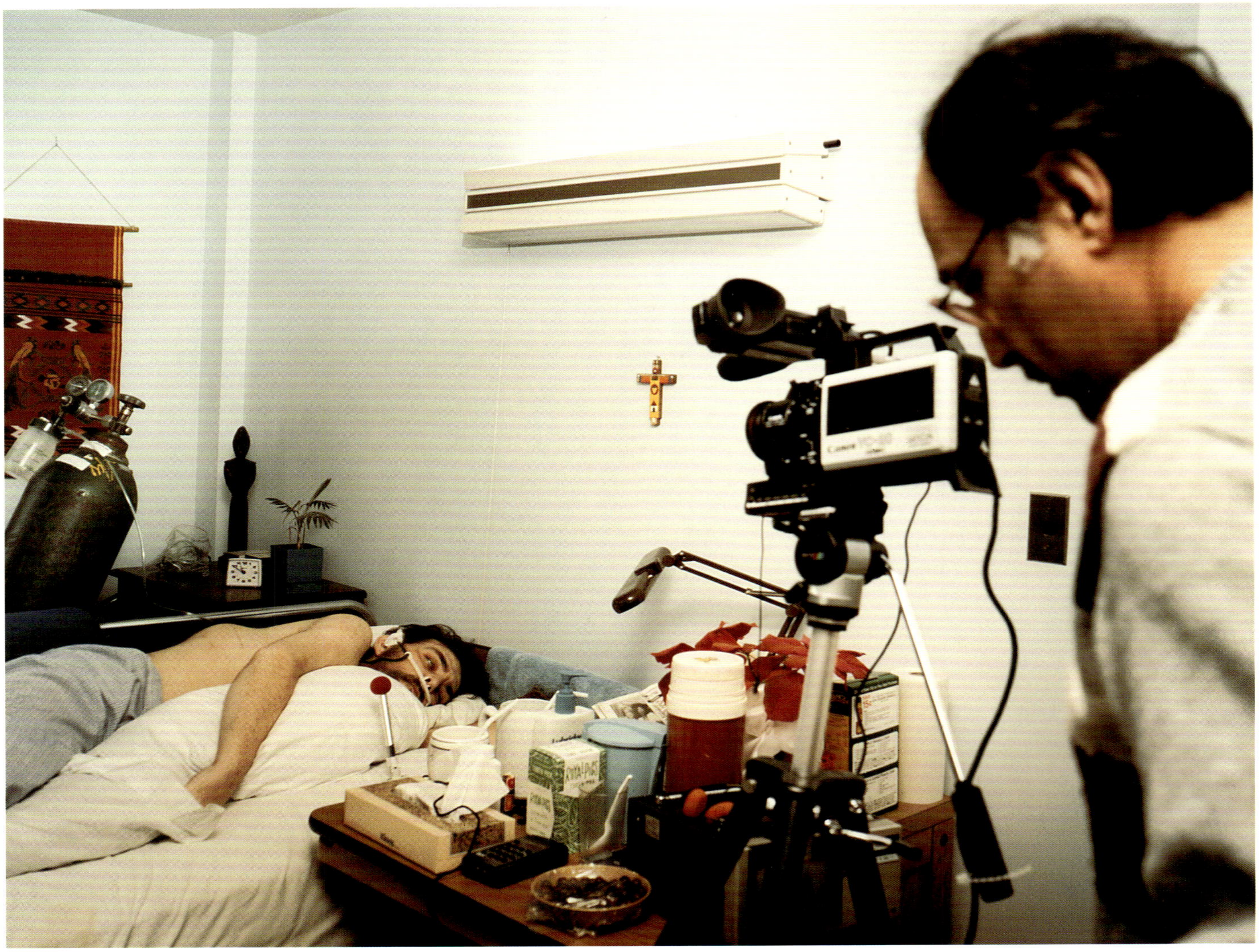

Father Richard Pankowski
Maryknoll Seminary, New York, 1986

Richard Pankowski was born in Manville, N.J., in 1950. In 1985, he was diagnosed as having pleural mesothelioma. This is an almost always malignant and invariably fatal tumor of the delicate membrane that encases the lungs, and it ultimately crushes the lungs so that the patient cannot breathe. It usually takes at least twenty five years to develop after the onset of exposure to asbestos, so Pankowski's exposure may well have begun when he was a child. At that time, mesothelioma was so rare that it was considered to be a pathological curiosity. It began turning up with increasing frequency, however, in people with some, even if slight, history of exposure to asbestos, such as family members of asbestos workers, people who lived close to asbestos mines or dumps, or people who worked only briefly with asbestos. Richard Pankowski's father worked at the Johns-Manville plant for more than 30 years. He often came home covered with asbestos dust, and died from asbestosis. While in college, Richard Pankowski worked at the Manville plant less than a year, unloading asbestos from trucks and boxcars, and cleaning machinery that was used in the manufacture of asbestos insulation. This photograph shows Father Pankowski as he is about to give a video-taped deposition for a lawsuit he had brought against Johns-Manville and other companies. At the time, he refused pain-killing medication so that he could be alert during cross-examination. The mesothelioma can be seen protruding through the rib cage on his right side. He died five months later.

Mrs. Carol Pankowski, 1986

Richard's mother, Carol, shown here (black dress) with the widows of two other asbestos workers, has also developed signs of asbestos disease. Her only known exposure to asbestos was to the dust her husband carried home on his work clothes.

(this page)
Johns-Manville Plant and Parking Lot
Manville, New Jersey, 1988

Oldest and largest (400-acre, 18 building tract) manufacturing asbestos plant in the United States. This site opened in 1912 and in 1985 was shut down.

(opposite page)
Clementine Szukis
Bridgewater, New Jersey, 1988

Disabled with asbestosis. She operated a beauty salon in the Manville area for more than 20 years. She was environmentally exposed to asbestos through her customers, who worked at the Johns-Manville Plant, New Jersey. She recalls women employees coming from JM down the road to have their hair done; she routinely would have to brush the asbestos dust out of those who worked as weavers in the textile division. Both her fingers and toes are clubbed—clubbing of the extremities is indicative of impaired blood oxygenation resulting from advanced lung scarring.

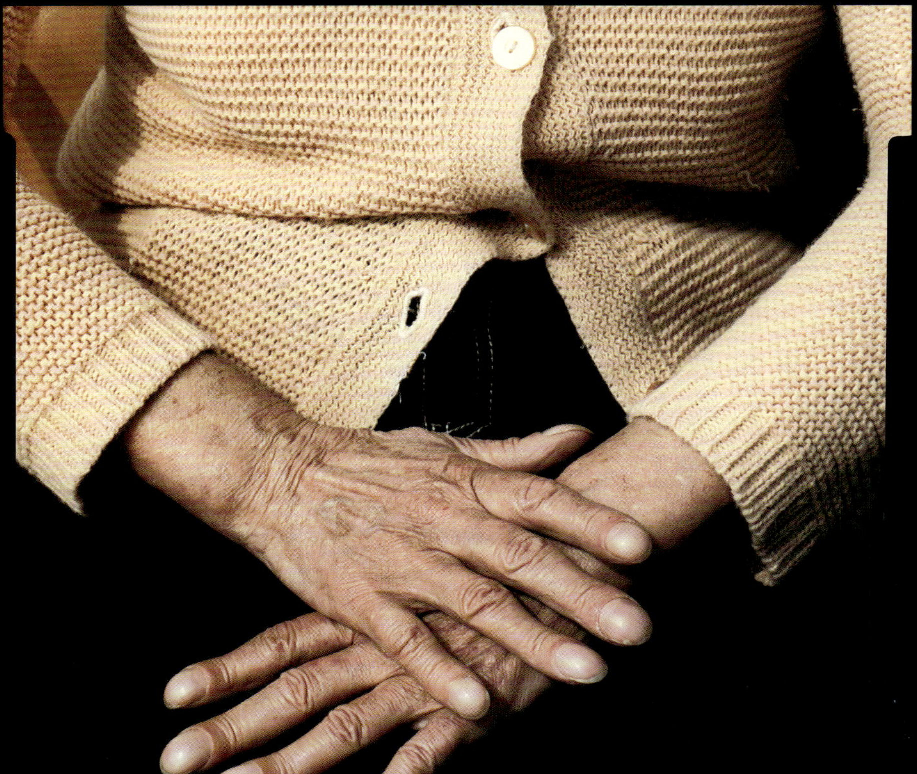

Ted and Betty Kowalski
Manville, New Jersey, 1988

Sixty-year-old Ted Kowalski was employed as a production worker at the Manville, N.J., plant from 1947 to 1967. He and his wife, Betty, who never worked at the plant, have been diagnosed as having developed asbestosis. So has their 37-year-old son, Teddy, who never worked at the plant, either. Here is how Ted and Betty look back on their lives:

Ted: *Years back, everybody that got sick with a lung problem was diagnosed as having TB, when actually it was asbestosis. But this is how they got away with it. They misdiagnosed you on purpose.*

Betty: *He used to go to work with dark working clothes. But then, by the time he would come home, he was like a snowman. The kids, Teddy and Donna, were small and they used to say, "Here comes daddy the snowman."*

Ted: *After I retired in 1966-67, my doctor treated me for chronic bronchitis up until 1980. Nowhere in my medical records does he state that I had asbestosis.*

Betty: *The saddest thing was when my son went. It was my birthday and he went to have himself checked. Ted went with him. When he came home, he got out of the car and he's walking down the sidewalk with his head down. I said, "Teddy, tell me that you don't have it, right?" He said, "Mom, I have it."*

Paul Safchuck
Baltimore, Maryland, 1989

President of the White Lung Association. Disabled with asbestosis.

As a consequence of man's utilization of asbestos, coupled with the natural occurence of the mineral, asbestos fibers are found in the air we breathe, the food we eat, and the water we drink.

1981 National Cancer Institute Study

Asbestos Road Pavement and Cinema
Asbestos, Quebec, 1988

(this page)
Tom Bowlby, September 1985

Mr. Bowlby, Bound Brook, N.J., worked at Johns-Manville Company, Manville, N.J., for more than three decades as a maintenance supervisor. At age 54 he was "awarded" disability and retired due to his asbestosis. Mr. Bowlby manifests one of the symptoms of this disease—finger clubbing.

(top, opposite page)
Tom Bowlby , September 1986

Mr. Bowlby and his wife Georgine at the Somerset Medical Center, Somerville, N.J. Mr. Bowlby has developed lung cancer from the asbestos exposure, which metastasized to the brain. Just prior to this hospital stay, he became mentally disoriented. Late one evening, he apparently crawled out of bed and attempted to bury himself in the backyard garden. His wife discovered him face down the following morning. He died a week after this photograph was taken at the age of 55.

(bottom, opposite page)
Tom Bowlby
Morgue, Somerset Medical Center, Somerset, N.J.
September 1986

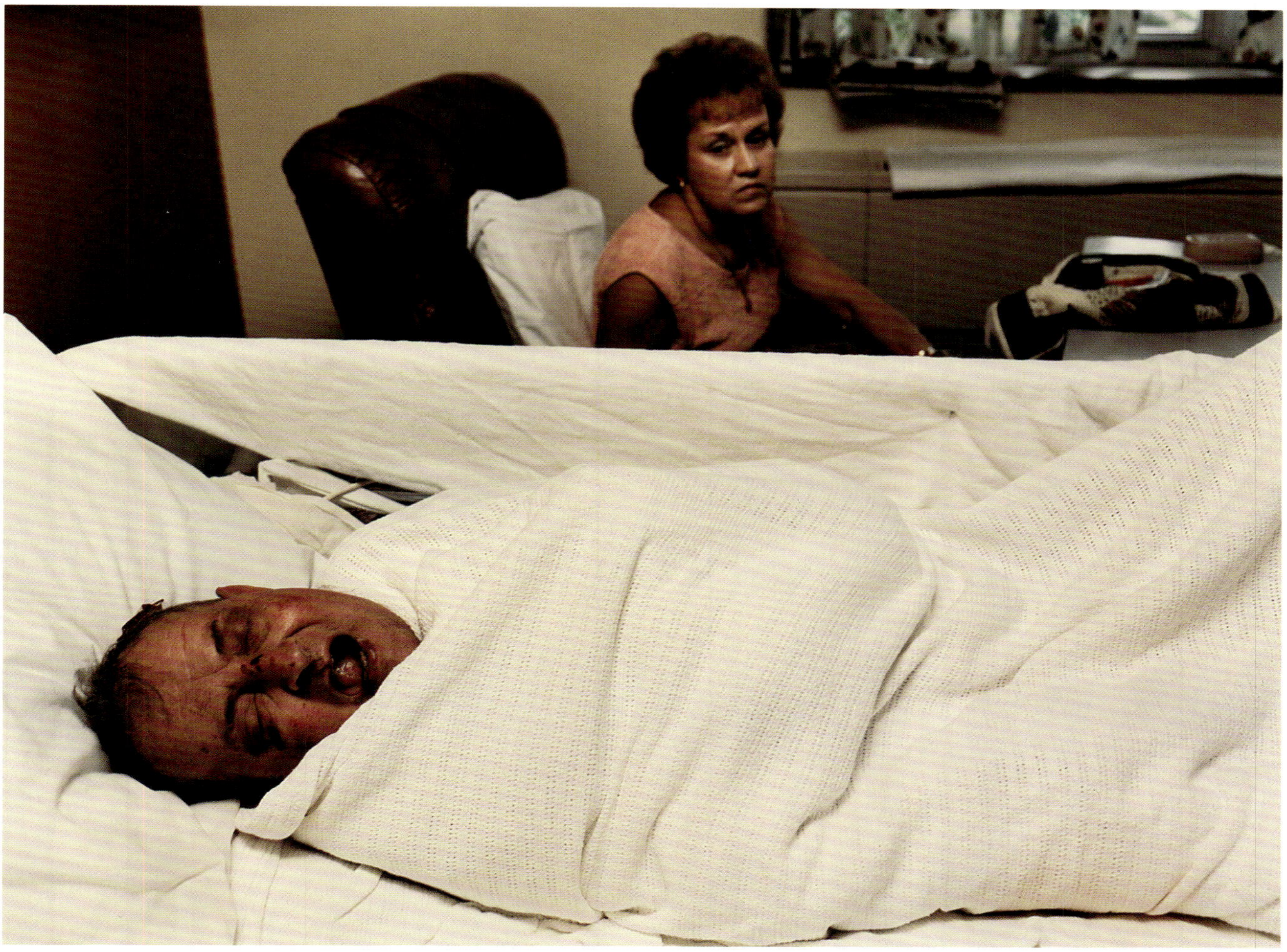

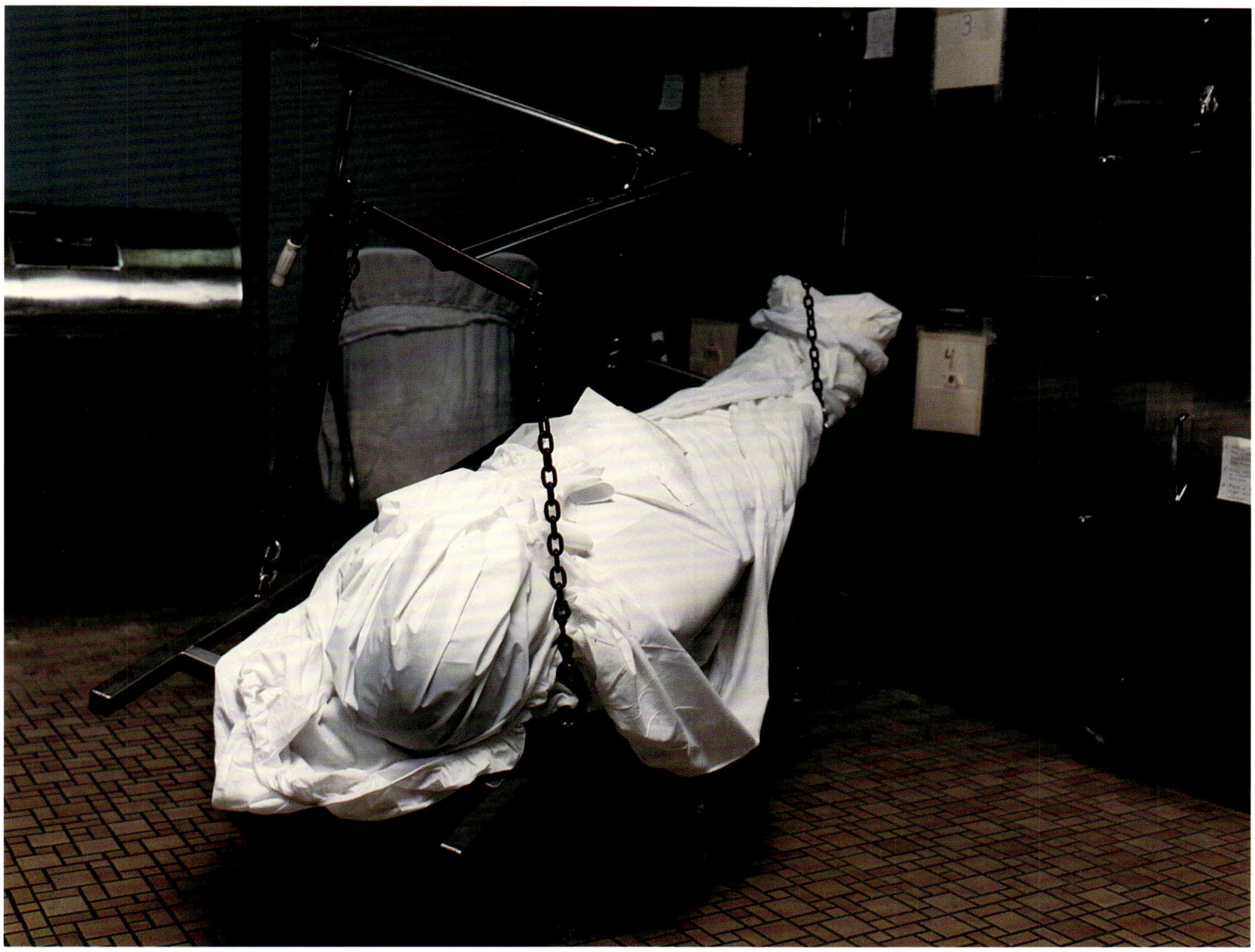

Jeffrey Mine, JM Asbestos
Asbestos, Quebec, 1988

The town of Asbestos is positioned around the rim of the pit which measures about a mile in length and 1,200 feet deep. The town has been moved several times over the last four decades to widen the mine site. This is the largest chrysotile asbestos site in the western world. Most of the asbestos sold from this site is being exported to third world countries. Last year the mine operated 364 days, on a double shift.

Workers with Dust
Johns-Manville Plant, Waukegan, Illinois, 1987

Boys Playing Caps
Manville, New Jersey, 1986

Most of the borough drinking water travels through transite pipe, which is made from a mixture of asbestos and cement. For the last several decades transite pipe was routinely installed in communities throughout America. The EPA ban/phase-down continues to allow transite pipe to be installed for drinking water pipe through 1996. More than 200,000 miles of asbestos-cement pipes are in use to carry drinking water to U.S. consumers. The pipes provide a source of asbestos fibers from both leaching and erosion. Studies have measured counts in the Los Angeles area of over a billion fibers per liter of tap water, by comparison, measurements in the state of Massachusetts, have a range of 1 to 44 million fibers per liter of municipal tap water.

The Third Wave Asbestos Conference: High Drama in Science

by David Kotelchuck, PhD, MPH

David Kotelchuck, a member of the Health/PAC Board, is Director of the graduate Environmental and Occupational Health Sciences Program at Hunter College of the City University of New York. He has written on asbestos science and policy for more than a decade.

Until recently, the scientific consensus that exposure to asbestos poses a serious risk to human health seemed well established and unlikely to change. But now a prestigious group of scientists has challenged one aspect of that consensus, creating controversy and turmoil among scientists and occupational health advocates.

In this context, the international conference on "The Third Wave of Asbestos Disease: Exposure to Asbestos in Place," held June 7–9, 1990, in New York City, provided a moment of high drama in the scientific community, with elements at once scientific, social, political, and personal. The conference was sponsored by the Collegium Ramazzini, an international body of environmental and occupational health scientists.

The Controversy Crystallizes

Since the early 1960s, following key articles by scientists such as Dr. Irving Selikoff,[1] now Professor Emeritus at Mt. Sinai Medical Center and President of the Collegium Ramazzini, and reinforced by hundreds of later studies, a scientific consensus has emerged and gathered powerful momentum that asbestos dusts are hazardous to human health, causing asbestosis, lung cancer, mesothelioma, and gastrointestinal cancers.[2] This in turn has helped shape a public consensus in the United States that asbestos dust and asbestos materials in place at work and in the community, especially in schools and other public buildings, must be carefully monitored and regulated.[3]

Among the remaining areas of scientific debate are whether certain crystalline forms of asbestos dust are more harmful than others, and thus should have lower permitted levels in the air. The federal Occupational Safety and Health Administration (OSHA) has a single exposure standard for all types of asbestos fibers, a position supported by Dr. Selikoff and many other U.S. scientists. However, the American Conference of Governmental Industrial Hygienists, an important private consensus group, recommends a standard for chrysotile asbestos fibers two-and-one-half times greater than that for crocidolite asbestos fibers.[4] Such a standard is in effect in England, reflecting the views of many British scientists that chrysotile is less dangerous than crocidolite. Notwithstanding this difference, both the U.S. and British standards regulate asbestos stringently.

Then, in January 1990, in an article in the prestigious *Science* magazine, several prominent scientists in the United States and Europe broke with the existing scientific consensus on asbestos.[5] They argued that chrysotile asbestos fibers, the type most commonly mined and used in North America, present a far less serious cancer risk than all other types of asbestos fibers (called amphibole fibers). Their contention was that many cancers among asbestos workers can be attributed to a 1 percent amphibole contamination in North American chrysotile deposits, and that in any case "data do not support the concept that low-level exposure to asbestos is a health hazard in buildings and schools."[6] The authors are prominent scientists at major academic institutions, including Dr. Morton Corn, former Director of the Occupational Safety and Health Administratoion under President Gerald Ford and now Director of Environmental Health Engineering at Johns Hopkins University; Dr. J. B. L. Gee, a pulmonary physician with the Yale University School of Medicine; and Dr. Brooke Mossman, a pathologist at the University of Vermont. The article's viewpoint was endorsed in an editorial by Dr. Philip Abelson, deputy editor of *Science*.[7]

While the *Science* article did not quite deny that chrysotile asbestos was dangerous, it clearly argued that its hazards were of a different order of magnitude than amphibole types of asbestos. Thus, it contended, laws like the federal Asbestos Hazard Emergency Response Act (AHERA) of 1986, which concerned asbestos in schools, far from being a reasonable response to the problem, instilled "fear" and "panic" among parents.[8]

The impact of the article on the scientific community was substantial. Occupational health scientists debated, at first informally, the article's sharp break with current thinking and the evidence for its so-called amphibole hypothesis. Also, the timetable for the New York City conference, some organizers have acknowledged, was moved up to allow prompt discussion of and response to these views.

The Social Impact

The social impact of the article was substantial as well, catching many supporters of asbestos workers and their families by surprise. We had heard about the so-called Harvard Asbestos Conference held in early 1989, a by-invitation-only private gathering of the asbestos industry and its legal and scientific allies—no Mt. Sinai scientists, for example, were invited.[9] Although the extent of Harvard's official participation was to rent the facilities on campus to the gathering, the conference was widely reported on as "the Harvard Conference," with the implied imprimatur of the university.

In part, the media response was the result of an aggressive, organized publicity campaign by the Safe Buildings Alliance, a publicity and lobbying arm of the asbestos building supply industry, including firms such as W. R. Grace, Celotex, and U.S. Gypsum.[10] The New York State Asbestos Advisory Committee, in its February 1990 report to Governor Mario Cuomo, accused the Safe Buildings Alliance of having conducted "an extensive and highly misleading disinformation campaign," including articles placed in *Readers Digest* and *Forbes* magazines.[11]

But behind the success of such a lobbying group was also a great demand from a variety of groups to hear this "good news." First and foremost, the North American asbestos industry benefits directly, since over 90 percent of the asbestos used in the United States is chrysotile fiber, the one type of asbestos fiber that is supposedly much less dangerous according to the amphibole hypothesis. Given the clouded history of asbestos research,[12] opponents of the asbestos industry have looked for—and found—associations with asbestos companies among scientists propounding the amphibole hypothesis.*

Also, the powerful real estate industry, suddenly stuck with vast amounts of empty office space and the gathering momentum of an economic recession, is desperately seeking to cut back on expensive asbestos removal and to fight proposed new state and local laws assessing and regulating asbestos in buildings. By April 1988, 38 states had passed laws regulating asbestos.[13] One such law is now moving toward passage in New York City, for example, where 84 percent of all large office buildings and 72 percent of all large apartment buildings contain asbestos, most of it in damaged condition.[14] These figures are probably not too much different from those in many other U.S. cities. Individual homeowners, who also face the health threats, emotional uncertainties, and financial risks of asbestos contamination in their homes are likewise a receptive audience for such "good news."

Of course, financially beleaguered state, county, and local officials, facing increasingly heavy asbestos-related expenditures during times of fiscal crisis, were quick to take note of the *Science* article and the relief implicit in its arguments. For example, during a March 1990 meeting with officials of the New York City Fire Department about possible asbestos awareness training, several of the officials asked this author whether I would speak about the "Harvard article." Obviously, these people had heard enough about the "Harvard" Conference and the *Science* article to confuse them, but otherwise had quite limited knowledge of the scientific literature in this field.

Besides attacking state and local asbestos laws, the real estate industry has attempted to influence public policy by organizing opposition to the federal AHERA law regarding asbestos in schools. The *Science* article feeds this opposition by asserting that AHERA instilled fear in parents, thereby encouraging haphazard asbestos removal, which in turn can lead to increases in airborne asbestos concentration, sometimes for months after removal.[15] Defenders of AHERA, such as the Service Employees International Union (SEIU), note that AHERA is essentially a "right-to-know" law, which simply requires schools to determine whether asbestos is present and develop a plan to prevent or control exposure for building employees and occupants.[16] Only in rare instances does the law mandate removal of the asbestos, but many building owners have been doing so routinely, not for compliance purposes but to recover the estimated 5 to 10 percent drop in resale value if a building contained asbestos.[17]

The Conference

In the midst of this controversy, 350 scientists, government officials, corporate and labor representatives, and lawyers from the United States and nine other countries, as well as 15 state and federal judges, attended the three-day Third-Wave Asbestos Conference. The conference co-chairs were Dr. Philip Landrigan of Mt. Sinai Medical Center and Dr. Homayoun Kazemi of Harvard Medical School.

The conference focused on the diseases caused by exposure to the 30 million tons of asbestos put into place in the United States since 1900. Many of those affected—building maintenance and repair, transportation, and other workers in public and private sector employment—are not normally considered asbestos workers, yet as the conference showed, they suffer in surprisingly large numbers from asbestos-related diseases. Earlier waves of asbestos deaths were associated with the mining and manufacture of asbestos products (the first wave) and the use of asbestos products among construction and shipyard workers (the second wave).

In his keynote address, Dr. Selikoff updated his continuing critical studies on mortality among asbestos construction workers of the second wave in the United States and Canada. His prospective study of 17,800 members of the asbestos insulation workers union, who were followed from January 1, 1967 to December 31, 1988, found that 4,951 had died when normally 3,453 would be expected to die—that is, 43 percent more workers died than expected! Four hundred and fifty-eight workers died of mesothelioma (9 percent), while no deaths would normally be expected. Thus, the death toll among asbestos workers continues, even as we debate the impact of asbestos exposure among those exposed to dust from asbestos in place.

Dr. William Nicholson, also of Mt. Sinai Medical Center, later presented a detailed analysis and ringing critique of the amphibole hypothesis. Central to his argument was a comparison of the results of nine epidemiological mortality studies of asbestos workers in which the amounts and types of fiber exposure were known. He found similar increases in lung cancer risk per unit of asbestos fiber exposure for all but one of the studies. The amphibole hypothesis would have predicted much lower cancer risks for several of the studies, those which involved primarily chrysotile exposure.

Nicholson also attacked as misleading some of the key evidence that Mossman and others had presented to support their arguments. They noted that upon autopsy, many victims of asbestos disease who had worked primarily with chrysotile showed elevated levels of the amphibole fiber tremolite, which is normally a 1 percent contaminant of North American chrysotile, and presented this as evidence that tremolite was causing the lung cancer. Nicholson asserted that this simply reflected the body's greater ability to break down chrysotile than tremolite fibers during the several decades it took the cancer to develop. It is a fallacious argument, he went on, to make inferences about the cause of a disease from events that took place decades later, at the end of the disease process.

In another key session on "Diseases Resulting from Asbestos Exposure in Buildings," three independent studies of asbestosis among building custodians were reported. Dr. Christine Oliver of Harvard Medical School reported that *33 percent* of 121 Boston school building custodians showed X-ray evidence of asbestosis lung disease. In New York City, Dr. Stephen Levin and colleagues at Mt. Sinai Medical Center studied 660 school custodians, members of the Operating Engineers Union, and found 28 percent with evidence of asbestosis. Among those in this group with 35 or more years on the job, *39 percent*—almost two of every five workers—had asbestosis. In Los Angeles, Dr. John Balmes of the University of California in San Francisco found 13 percent of the custodians studied showed evidence of asbestosis. Since these custodians all worked in U.S. buildings, which therefore contained primarily chrysotile fibers, and since all the cases involved incidental exposures in school settings, the high incidence of asbestosis in this group flatly contradicts the *Science* article's assertion that "available data do not support the concept that low-level exposure to asbestos is a health hazard in buildings and schools."[18]

These reports were followed by an epidemiological study of the occupations of 254 mesothelioma cancer mortality cases in Wisconsin, drawn from the state's cancer registry, by Dr. Henry Anderson and associates. As expected, they found the greatest likelihood of developing mesothelioma among insulation workers and shipbuilders. But of particular interest were statistically significant elevations in the odds of developing the disease among "non-asbestos workers," including fire fighters, law enforcement officers, school employees, postal workers, and janitors. These workers have in common offices and employment in public buildings, many of which are contaminated with asbestos dust.

More than 50 other studies were reported at the conference, the full proceedings of which will be available soon from the New York Academy of Sciences. What was most persuasive was not any individual report, but the cumulative impact of report after report describing worker groups and some communities around the world struck by asbestos disease. Many of those reporting from countries outside the United States were not only prominent scientists in their native lands, but important public health officials there as well. And if many had been slow to overcome their initial skepticism about the extent of asbestos-related diseases during the 1960s and 1970s, they are now clearly convinced of the health hazards of asbestos and are gathering strong evidence to buttress their views.

They heard reports of major new findings on the effects of asbestos in schools, as well as new results that extended the boundaries of occupations affected by asbestos exposures and communities affected by dust from asbestos mineral outcroppings. These results vault over the artificial distinctions constructed in a recent *New York Times Magazine* article on the conference that "No new evidence was presented to demonstrate that deteriorating asbestos in schools and public buildings was causing increased cancer rates in the general population. However, a mountain of papers were presented that detailed the health effects of asbestos exposure among workers."[19] (The *Times* article also failed to understand the scientific import of the Nicholson analysis and other critiques of the amphibole hypothesis—they are not even mentioned in the article—although the hypothesis is based essentially on another reexamination of existing data, with little or no new data.)

In the end, the Third-Wave Asbestos Conference refuted the key assertions of the *Science* article with more and stronger scientific studies. Such good science, admirable as it is, will not alone triumph over poor or faulty science, any more than the truth alone will set us free. But without the scientific truth about asbestos, we cannot be free of the moneyed, virulent self-interest of the asbestos industry or the deaths and personal devastation that follow in its wake.

* *"Dr. Mossman, affiliated with the University of Vermont, consults with Owens Corning. Recently, she withdrew her name from a defense witness list to avoid a subpoena seeking to take testimony about the* New England Journal of Medicine *article. (Ironically, the University of Vermont is removing asbestos from its offices and has initiated litigation against the asbestos companies to recover the cost. Obviously, the University either hasn't read or doesn't believe Dr. Mossman's research.) Dr. Gee, who is a well-known defense witness in the Boston area, has stated in depositions that he spends more than 70 percent of his legal and medical consulting activities on behalf of defendant asbestos companies. Dr. Corn, a strident advocate for defendant asbestos companies, testified in a recent deposition that he destroyed his file on the* Science *magazine article so building owner attorneys could not see what was in it. Dr. Corn received more*

than half of his outside income from asbestos companies last year."—Testimony of the Service Employees International Union before the Subcommittee on Health and Safety, Committee on Education and Labor, U.S. House of Representatives, April 4, 1990.

References

1. Selikoff, I.J., J.Churg, and E. C. Hammond, "Asbestos Exposure and Neoplasia," *Journal of the American Medical Association,* April 6, 1964:188, pp. 22-26; and Selikoff, Churg, and Hammond, "Relation between Exposure to Asbestos and Mesothelioma," *New England Journal of Medicine*, 1965:272, pp. 560-565.

2. National Institute for Occupational Safety and Health, *Workplace Exposure to Asbestos: Review and Recommendations*, Pub. No. 81-103, Washington, DC: U.S. Department of Health and Human Services, 1980; and Occupational Safety and Health Administration, "Occupational Exposure to Asbestos, Tremolite, Anthophyllite and Actinolite" (preamble), *Federal Register*, June 20, 1986:51(119), pp. 22615-22650.

3. National Emission Standard for Hazardous Air Pollutants, 40 CFR Part 61, Subpart M (1973); Occupational Safety and Health Administration Asbestos Standard, 29 CFR Part 1910, Subpart Z (1986); and Asbestos Hazard Emergency Response Act, 40 CFR Part 763 (1986).

4. *Threshold Limit Values and Biological Exposure Indices for 1989-1990,* Cincinnati, OH:American Conference of Governmental Industrial Hygienists, 1989, p. 12.

5. Mossman, B.T., J. Bignon, M. Corn, A. Seaton, and J.B. L. Gee, "Asbestos: Scientific Developments and Implications for Public Policy," *Science*, January 19, 1990:247, pp. 294-301.

6. Mossman et al., op.cit., p.294.

7. Abelson, Philip, "The Asbestos Removal Fiasco," Editorial, *Science,* March 2, 1990:247, p. 1017.

8. Mossman, et al., op.cit., p. 299.

9. Interview with Irving J. Selikoff in *Asbestos Action*, newsletter of the White Lung Association of New Jersey, May 1990, p.3.

10. *Safety Rep*, newsletter of the New York Committee for Occupational Safety and Health (NYCOSH), Fall 1990, p. 8.

11. State of New York Asbestos Advisory Board, *Second Annual Report,* February 1990.

12. Kotelchuck, David, "Asbestos: The Funeral Dress of Kings—and Others," in D. Rosner and G. Markowitz, eds., *Dying for Work: Workers' Safety and Health in Twentieth-Century America,* Bloomington: Indiana University Press, 1987, chapter 12; and Castleman, Barry I., *Asbestos: Medical and Legal Aspects*, 2nd ed., New York: Harcourt Brace Jovanovich, 1986.

13."Buildings with Asbestos Shunned," *New York Times*, April 28, 1988, p. D1.

14. *Final Report on the Assessment of the Public's Risk of Exposure to In-Place Asbestos.* New York City Department of Environmental Protection, 1988, reported in *Safety Rep*, Fall 1990, p.6.

15. Mossman et al., op cit., p.299.

16. Service Employees International Union, Testimony before the U.S. House of Representatives Committee on Education and Labor, April 4, 1990.

17."Buildings with Asbestos Shunned."

18. Mossman et al., op. cit., p. 294.

19. Hooper, Joseph, "The Asbestos Mess," *New York Times Magazine*, November 25, 1990, p.53.

Canada's Growth Market: Asbestos Cancer in the Third World

Barry I. Castleman, ScD

Barry Castleman is an environmentalist and researcher specializing in health issues. Castleman is the author of Asbestos: Medical and Legal Aspects, *which is the standard reference for understanding the history of asbestos-caused diseases. His current work addresses the dumping of asbestos in developing countries.*

Canada is the world's leading exporter of asbestos. In the past, most of the mineral was sent to the U. S. and other industrial countries. With growing awareness of the hazards of asbestos and regulation in these countries, their imports of asbestos have plummeted. Over 100,000 people in the U. S. have died from asbestosis and cancer from asbestos. The cancer toll from asbestos in the U. S. throughout the 1990s will exceed one death per hour. The economic costs to our society will be hundreds of billions of dollars.

Canada is now looking to the developing countries as the market for asbestos in the 1990s. Unfortunately, conditions of current asbestos use in Asia, Africa, and Latin America resemble those that existed here before workers learned of the time-bomb effects of breathing asbestos dust. This exploitation of ignorance and poverty for a quick buck has the strong backing of the government of Canada, even though Canadian workers would walk off the job under the conditions of asbestos use in the poor countries. What began as a commercial lie through most of this century has thus become an official lie. Canada's Asbestos Institute proclaims, "Implementing safety in the use of asbestos is neither difficult nor expensive."

The 1990s will see growing international opposition over the export of asbestos and its use in the poor countries. Wood pulp and other safer substitutes already are being used to make fiber-cement panels and roofing, the most visible market for asbestos in the Third World.

Introduction

In the technological development of nations, it has been widely accepted that each country would do well to choose "appropriate technology" based on its resources, its people and their needs. Public health impacts of some technologies render them undesirable for Third World country development.

For example, no one would urge that developing countries should build plants to make asbestos thermal insulation. Asbestos has been replaced by wood pulp, fibrous glass, and other materials to make insulation products far less deadly than the asbestos insulations used in the past. Thermal insulation reinforced with asbestos is an example of *discredited technology* because of its severe, unavoidable hazards and the availability of safer alternatives. The most recent manufacture of this product I know of was by Johns-Manville in Brazil in 1980; however, it may still be made by producers in Thailand and/or India.

Industry Theory: Controlled Use Exists and Is Becoming Universal

The international asbestos industry has claimed for decades that asbestos can be used safely and that it should be used in Third World development. Corrugated asbestos-cement roofing is used worldwide, and together with other asbestos-cement sheet and pipe products accounts for over 80 percent of asbestos use. Increased use of asbestos in developing countries has offset tremendous market losses in the industrial nations over the last 10 years. Quarterly publications of the Asbestos Institute in Quebec describe aggressive sales efforts in Asia, Africa, and Latin America.

The International Labor Office has published a "Code of Practice" enumerating very basic safeguards that should be used in work with asbestos.[1] Some countries have regulations that apply to industries where asbestos hazards exist. The issuance of control instruments as published documents, however, never has and never will assure that the "mandated" controls are in fact applied. The gulf between what is advertised as "controlled use of asbestos" and the reality of manufacturing and construction work with asbestos is greatest in the poor countries whose use of asbestos is on the rise.

Widespread Uncontrolled Use in Developing Countries

The chief of Brazil's environmental protection agency wrote in 1986 that the labor authorities in charge of worker protection did "poor work" and were "very ineffective."[2] As of 1986, this official wrote that, "we don't make any [power tools with exhaust ventilation] in Brazil, and it is difficult to import them."[3] It seems highly unlikely that portable saws with exhaust ventilation and dust capture are being used by construction workers handling asbestos-cement products in countries like Brazil. When that country's authorities began to inquire about health risks in asbestos manufacturing operations in 1980, the government people depended upon companies visited to provide and demonstrate the use of standard air monitoring equipment. As of 1986, the official workplace exposure limit for asbestos in Brazil was 4 fibers/cubic centimeter, twenty times as high as the 0.2 fibers/cc limit in the U.S. (Even at the U. S. limit, which is now slated to be slashed again by half, workers inhale millions of asbestos fibers per day, and occupational cancer is expected to cause death to 6/1000 of the workforce exposed.)

Uncontrolled use of asbestos has been the norm in many countries in recent years. Brazil's asbestos industry employs about 30,000 workers, most of them in small plants, and conditions are so bad that asbestosis is being diagnosed 5-10 years after the workers enter the industry. Mexican researchers found severe asbestosis in workers employed spraying asbestos, who were monitored as having exposures of 54 fibers/cc in *1982*.[4] Sprayed asbestos products have been recognized as a mortal hazard from the time they were introduced in 1932. Investigations in India showed complete disregard for worker health by affiliates of U.S. and British multinational corporations making asbestos products.[5] In one Indian plant, where I have been told asbestos-cement pipe was sawed without local exhaust ventilation, government hygienists measured exposures of 216-418 fibers/cc.[6] In Egypt, workers apparently still empty asbestos bales and mix asbestos with cement manually.[7]

If the asbestos industry is taking concerted action to implement "controlled" use of asbestos today, it represents a complete reversal of recent practices. In 1977, Canadian asbestos mining firms arranged to delete warning labelling about the cancer hazard of asbestos, opting instead to accept written releases of liability from a distributor in Japan.[8] Similarly, 1978 minutes of the Asbestos International Association reveal an international conspiracy to proceed as slowly as possible, country by country, using the weakest possible warning labels "in fear of a possible influence on sales."[8,9]

Given the historic lack of both industry product stewardship and controlled asbestos exposures, especially in the vulnerable developing countries, the operative question is: *Will* asbestos hazards be controlled? (not: *Can* asbestos hazards be controlled?) The burden is on the asbestos industry to demonstrate that it is practical to routinely use asbestos in a thoroughly controlled way in developing countries.

The record to date suggests that it is unreasonable to expect that asbestos hazards will be controlled in the developing countries. Industry spokesmen acknowledge that, even now, construction contractors in the U.S. sometimes use abrasive disc saws to cut asbestos-cement pipe — despite advice against the practice by the A-C Pipe Producers Association and the existence of applicable Occupational Safety and Health Administration standards since 1972. Lung disease has occurred in the U.S. from the uncontrolled sawing of asbestos-cement sheets in construction work in the 1970s.[10]

How then can we expect Third World manufacturers of asbestos products and construction contractors to take on the cost of extraordinary control measures, when there typically isn't even pressure from industry or government authorities to do so? Where is the infrastructure of prevention (information, regulation, and compensation) in Third World countries? And why should developing countries submit to the likelihood of asbestos contamination and disease, when safer alternatives exist that will not warrant the unprecedented commitment of scarce public health resources?

Developing countries may also wish to consider another form of "pollution" that has frequently come along with the growth of an indigenous asbestos industry. This is the corruption of the fledgling professions of industrial medicine and hygiene, as pressures are brought to bear on health professionals in industry, government and academia to learn the "industry line" and downplay concerns about workplace and environmental exposure to asbestos. This impact on a vital sector of a society in development may pave the way for subsequent public health abuses in other industries. Again one must wonder why a developing country would want to accept the externalized costs of a growing asbestos industry, given the alternatives now available.

Asbestos Substitutes

In the 1980s, an increasing array of asbestos-free products has become commercially available. Asbestos-containing corrugated and flat cement sheeting, valve and pump packings, roofing felts, pipeline wrap, and vinyl flooring have not been made in the United States since the mid-1980s. In Europe, Nueva, A.G. agreed to eliminate asbestos in fiber-cement sheet products in 1990 in Germany and Switzerland, in favor of toxicologically-tested plastic-fiber reinforced cement sheet. The Swiss government has allowed until 1995 for asbestos use to be discontinued in making fiber-cement pressure pipes.

Nueva has already been able to replace asbestos with wood fiber and sisal substitutes in Costa Rica and other Latin American countries.[11] Nueva's asbestos-free products carry warranties equal to those of the predecessor asbestos products and are made with locally available trees. Nueva does not plan to compete with cheap domestic asbestos in Brazil and Colombia or imported asbestos in Mexico, unfortunately. In Australia and Malaysia, James Hardie Industries and its affiliates are also making cement sheet products reinforced with wood pulp instead of asbestos.

This is progress indeed, when one considers that exposures from sawing asbestos-cement sheet without dust controls have been reported as hundreds of fibers per cc.[12] It is also relevant that in many countries people catch water running off their roofs for drinking and cooking. Environmental Protection Agency researchers have reported that "asbestos fiber concentrations over 500 million fibers per liter have been found in cistern drinking waters which use asbestos-cement roofing tiles to collect water."[13] It would be better if people had this burden of wood, coconut, or banana plant materials instead of asbestos in their drinking water.

The Role of the Canadian Government in Promoting Asbestos Use

When the U.S. Environmental Protection Agency proposed to ban asbestos, the Canadian government, representing both private and state-owned asbestos mines, applied considerable pressure to oppose the ban.[14] An article in the British magazine *The Economist* created a furor, for it suggested that Canada had become "a sort of merchant of death by unloading its asbestos on unsophisticated Third World clients who may not be aware of its dangers."[14,15] Canada's Energy, Mines and Resources Minister Marcel Masse in 1988 responded to the above article by writing, "[t]he risk can be managed anywhere. This includes the Third World, where governments are more aware of the risks and more capable of controlling them than your correspondent is willing to credit."[15] *The Toronto Star* went on to describe a $30 million campaign of federal and Quebec governments to "try to drag asbestos out of the doghouse."[15]

The Asbestos Institute, which is partly supported by Canadian taxpayers, also worried that EPA's asbestos "ban", proposed in 1986 and issued in 1989, would impede the promotion of asbestos in countries which have little or no experience, let alone resources, in controlling industrial cancer threats. The Institute is a joint venture of the asbestos mining industry and the government, "to maximize the use of existing resources in a concerted effort to defend and promote the safe use of asbestos on a global scale." It claims to be "dedicated to promoting the proper use of asbestos."[16]

Canadian physician David Bates has called for the establishment of an independent commission ("recruited neither from industry nor from government employees") to monitor certain indicators of product stewardship in the export of Canadian asbestos and report annually to the public.[17] An appropriate topic for such an oversight panel would be the publication of lies like this by Asbestos Institute President Claude Forget: "In [Selikoff's] study of American insulation workers, asbestosis victims did not only inhale white asbestos as you state but *were exposed to mostly amosite asbestos.*"[18] Or this, from Asbestos Institute's "The Real Facts on Asbestos" (1990): "[Third World] construction sites, however, tend not to be a major problem as hand-operated tools, which in general generate large particulate dust and not respirable dust, are widely used." Hand sawing asbestos-cement panels generates respirable fiber counts from 31 to over 200 fibers/cc, according to an earlier industry report.

Canada's independent asbestos oversight panel, if it is ever set up, might also want to monitor the intimidation and vilification of scientists at conferences where the hazards of asbestos are discussed. For example, there is this by the Asbestos Institute Director of the Health and Environment Division, Jacques Dunnigan, at an asbestos conference in Mexico: "It is very hard for me to abstain from expressing my feeling that what we have just heard is standard, usual, ad nauseam repeated practice of some people at Mt. Sinai."[19] One doctor's repeated efforts to inject obfuscation into the proceedings during a major asbestos conference in New York in 1990 led many to conclude that he was a paid heckler.

Dr. Bates also called on the government of Canada to provide as much money for research into asbestos' health effects as is released for promotion of the industry. This balance, along with the creation of the oversight commission "would help to reassure average Canadians that they could not be accused of simply cynical exploitation of other people's ignorance," he concluded.[17]

Canada's union of insulation workers and other unions feel it *is* unworthy for the Canadian people to be represented by government officials who would rather sell ever more asbestos than plan for an asbestos-free future. Canada has spent over $500 million on post-1978 government takeovers and subsidies to bail out its asbestos mining industry. Closing of this "poisonous" industry with its 3000 jobs has been compared to the closing of an unneeded military base by Dr. Ray Sentes, a former insulation worker with asbestosis.[20] Canadians have also been urged to close the mines by Dr. Irving Selikoff of the Mt. Sinai School of Medicine in New York, the physician whose research on asbestos workers and courage to speak out have fueled public health efforts worldwide for over 25 years.

Nationally-viewed television documentaries in Canada (CBC, Oct. 1988; CTV, Nov. 1989) critically examined the government's practice of pushing asbestos in the Third World. These programs included vivid filming of uncontrolled asbestos use in construction in Thailand and the Philippines.

Canada's decision to sue the U.S. Environmental Protection Agency in a precedent-setting case, over EPA's ban/phase-down rule issued in 1989, was also criticized. In the CTV program, the Asbestos Institute spokesman was visibly upset to hear that the new Minister of Energy, Mines and Resources had declined to be interviewed about asbestos. It is unlikely that this signalled a change in government policy, however, and seems to have been simply the extraordinary arrogance of a political leader.

The Canadian government's support of the asbestos industry in the lawsuit in the U.S. is indeed unfortunate. The EPA "ban" will not affect any U.S. industry now using asbestos as a raw material until 1994. At the rate the industry is declining, EPA's regulation may amount to nothing more than closing the coffin after the banned industry is already dead. Moreover, Canada's suit against EPA adds little or nothing of substance to the issues the asbestos industry could raise by itself, based on the rulemaking record. In other words, the good name of Canada is being dragged through the mud, so that country's taxpayers can defray $1 million or so of the legal fees that would otherwise be borne by the international asbestos industry.

Canada has displayed an inconsistent and opportunistic regard for the U.S. legal system. Canadian asbestos mining interests have repeatedly refused to honor U.S. courts' evidence requests and verdicts awarding damages for failure to exert adequate product stewardship in the past (no sacks of asbestos were shipped from Canada with any health warning labels until 1969). Yet even as the toll of asbestos victims mounts, publicly and privately held mining interests in Canada want to be forgiven their debt to past consumers of their product while being entitled to continue creating future asbestos victims in the U.S.

It is tragic that Canada has lowered itself to become a litigant in U.S. courts, only to come in on the wrong side of a public health issue. Had Canada instead sued the EPA over the Reagan Administration's protracted refusal to regulate the sources of acid rain, Canada's use of U.S. courts would have been welcomed on both sides of the border by environmentalists and public health workers.

The Soviet Union's performance as a major exporter of asbestos fiber and technology is also worthy of scrutiny from a health standpoint. Reliable information on this would be most welcome, whether from the USSR, Canada, or other sources.

Conclusion

The eventual elimination of asbestos in favor of safer (and in some cases essentially harmless) materials is of great public health importance. The continued lack of product stewardship by companies and countries mining asbestos, in the interim, constitutes a formidable health threat. It can only be hoped that this industry will see that its credibility and near-term survival depend upon worldwide implementation of unprecedented controls. Some countries may still decide that, at least for now, the continued use of asbestos-cement water conduits, for example, is still justified in health and economic terms. But even without the demonstration that asbestos *will* be used with stringent safeguards, the industry is sure to suffer rapid declines as social opposition mounts. Anyone who doubts the swiftness with which such events can move should recall West Africa's revulsion and response to being used as a dumping ground for hazardous wastes from the U.S. and Europe in 1988-89.

My own experience with the asbestos industry leads to the conclusion that the only way to assure that asbestos will stop killing people needlessly is to ban it. This approach, which has been taken in Sweden and the United States, is even more attractive in developing countries where stringent regulation is not really a viable alternative to a ban.

References

1. Atherley, G. *et al.*: *Safety in the Use of Asbestos*. International Labor Office, Geneva (1984).
2. Noguiera-Neto, P.: letter to B. Castleman (Feb. 6, 1986).
3. Noguiera-Neto, P.: letter to B. Castleman (Mar. 12, 1986).
4. Mendes Vargas, M.M., Maldonado Torres, L.: Aplicacion de Asbesto por Aspersion y Neumoconiosis muy Incapacitante. *Revisita Med.* 20:73-77 (1982).
5. Castleman, B.I. : Double Standards: Asbestos in India. *New Scientist* 89: 522-23 (1981), Also, Castleman, B.I., Vera Vera, M.J.: Impending Proliferation of Asbestos. *Internat. J. Health Serv.* 10: 389-403 (1980).
6. *Asbestos: the Dust that Kills*. Centre for Education and Documentation (Bombay), p. 22 (1983).
7. Kamal, A.A.M. *et al.*: Usage of Personal Protective Devices among Egyptian Industrial Workers. *Am. J. Indust. Med.* 13:706-716 (1988).
8. Castleman, B.I.: *Asbestos: Medical and Legal Aspects*, 3rd Ed., Prentice-Hall Law & Business, Englewood Cliffs, NJ (1990).
9. Asbestos International Association: 8th Exec. Committee Mtg., Agenda Item 6: Labelling (July 7, 1978).
10. Harless, K.W., Watanabe, S., Renzetti, A.D.: The Acute Effects of Chrysotile Asbestos on Lung Function. *Env. Res.* 16: 360-72 (1978).
11. Comments of the Swiss Eternit Group on the EPA's Notice of Proposed Rulemaking on Asbestos, Washington (June 28, 1986).
12. Cross, A.: Practical Methods for the Protection of Men Working with Asbestos Materials in Shipyards. *Safety and Health in Shipbuilding and Ship Repairing*, pp. 93-101. International Labor Office, Geneva (1972).
13. Millette, J.R. *et al.* : The Need to Control Asbestos Fibers in Potable Water Supply Systems. *Sci. Tot. Env.* 18: 91-102 (1981).
14. Canadian Asbestos/Export and Die? *The Economist*. pp. 82-83 (Sept. 26, 1987).
15. Cahill, J.: A $30 Million Campaign Aims to Take the Curse Off Asbestos. *Toronto Star* (Mar. 7, 1988).
16. Asbestos. Vol. 1 (No. 1). Asbestos Institute, Quebec (1985).
17. Bates, D.V.: Asbestos: Promotion or Prohibition? *Can. Med. Assoc. J.* 136: 107-109 (1987).
18. Forget, C.: Dusty Reply. *The Economist*. p. 8 (Dec. 12, 1987).
19. *Proceedings, Meeting on Asbestos and Health in Latin America*, pp. 74-75. (M. Mitastein, Ed.) Pan American Health Org., Mexico City (1987).
20. Sentes, R. Canada's Hazardous Defense of Asbestos. *Financial Times of Canada*, p. 41 (Sept. 25, 1989).

Prepared for the VIIth International Pneumoconioses Conference, Pittsburgh, August 23-26, 1988. Revised Dec. 19, 1990.

Victims' Groups

Asbestos Victims of America
P.O. Box 559
Ste. L
Capitola, CA 95010
(408) 476-3646

Asbestos Victims Special Fund Trust
1500 Walnut St.
Mezzanine Floor
Philadelphia, PA 19102
(800) 447-7590

Asbestos Victims Education and Information
312 Essex St.
Gloucester City, NJ 08030
(609) 456-5695

Association to Improve Benefits for Retired
Disabled Workers & Their Dependents
27 N. Eleventh St.
Manville, NJ 08835
(908) 526-0679

White Lung Association
National Headquarters
P.O. Box 1483
Baltimore, MD 21203-1483
(301) 243-5864
(410) Area Code Effective 11/91

White Lung Association of New Jersey
901 Broad St.
2nd Floor
Newark, NJ 07102
(800) 395-9565

White Lung Asbestos Information Center
45 West 10th St.
Apt. 2G
New York, NY 10011
(212) 777-1882

Selected Bibliography on The Asbestos Project

Reviews

Bale, Tony. "Asbestos in Focus," editorial, *Health/ PAC Bulletin,* Vol. 20, Number 4, New York City. January 1991.

Chadwick, Susan. "Ravages of Asbestos Captured in Photos," *The Houston Post*, Houston, TX. May 26, 1990.

Brugge, Douglas. "Breath Taken, *Unity*, " San Francisco, CA. May 1990.

Wise, Kelly. " 'Breath Taken': Chilling, striking exhibit," *Boston Globe*, Boston, MA. April 3, 1990.

Bennett, Jonathan. "Asbestos cuts swath of death through industry," *The Guardian*, Vol. 42, Number 22. March 28, 1990.

French, Desiree. "Haunting look at the effects of asbestos," *Boston Globe*, Boston, MA. March 1, 1990.

Recinto, Ron. "Symposium and exhibit trace effects of asbestos exposure," *The Daily Free Press,* Boston, MA. March 13, 1990.

Feinberg, Jody. "Ills of asbestos aired in exhibit, symposium," *Patriot Ledger*, Quincy, MA. February 28, 1990.

Vitez, Michael. "His focus is on the deadly face of asbestos," P*hiladelphia Inquirer*, Philadelphia, PA. April 13, 1987.

Feature Articles, *photographs reproduced*

Socolar, Debbie. "Breath Taken, Exposing the Ongoing Tragedy of Asbestos," *Health/PAC Bulletin*, Vol. 20, Number 4, Cover Story and Portfolio, New York City. January 1991.

Johnston, Patricia. "Breath Taken: The Landscape & Biography of Asbestos—An Interview with Bill Ravanesi," *Views, The Journal of Photography in New England,* Photographic Resource Center, Vol. 11, Number 2, Boston, MA. Winter 1990.

Sentes, Ray. "Asbestos: The Silent Killer Lives On" *Canadian Dimension*, Cover Story. Winnipeg, Canada. May/June 1990.

Marantz Henig, Robin. "ASBESTOS: Its Toll on the Body and the Spirit," *The Washington Post Magazine Health Section,* Cover Story and Portfolio, Washington, DC. May 19, 1987.

Broadcast Reviews

WGBH-TV, PBS (Boston), 10 O'Clock News. 8-minute segment on the *Breath Taken* exhibition. Aired March 2, 1990.

WBUR, NPR Radio (Boston), Morning Edition, 6-minute segment reviewing the *Breath Taken* exhibition. Aired March 1, 1990.

Breath Taken, a 36-minute video produced and directed by Ravanesi, was cited for an honorable mention award at the 38th Annual Columbus, Ohio, 1990 Film and Video Festival in the documentary genre, and selected for screening in the upcoming 1991 U.S. Environmental Film Festival.

Biographical Notes on Bill Ravanesi

Breath Taken is the most recent in a series of long-term documentary projects that address social and cultural issues and on which Ravanesi has been working since 1974. He has established a reputation as part of the New Color movement in contemporary photography and is also known as a photographer of working class America.

For several years, he photographed farm labor in the South and Southwest. That series concentrated on Filipino migrants and included portraits of farm laborers and agricultural landscapes. *Workers on the Land: South/Southwest,* travelled nationally under the auspices of the Ford Foundation between 1981 and 1987.

His subsequent interest in inner city life in the United States culminated with the Holyoke Public History Project. That series focused on the city's industrial history and ethnic labor force, with particular emphasis on the last wave of Puerto Rican workers. It culminated with the exhibition, *The Hidden Holyoke: A Cityscape and Its People.*

Ravanesi has exhibited extensively in both the United States and abroad. He has received fellowships and grants from the National Endowment for the Arts, the Ford Foundation, the Massachusetts Artist Foundation, the Massachusetts Council on the Arts and Humanities, the Massachusetts Foundation for Humanities and Public Policy, and the Asbestos Victims Special Fund Trust, among others. He has also held teaching fellowships at Drexel University and Amherst College.

His undergraduate degree was awarded in biology from the University of Massachusetts in 1969. From 1972 and 1975 he attended Imageworks: The Center for Visual Studies, in Cambridge, Massachusetts and in 1978 he received a master's degree in photography from Goddard College in Vermont.